The Science of

$$t' = \frac{t}{\sqrt{1 - \frac{v^2}{c^2}}}$$

$$\vec{F_g} = \frac{G\,m_1}{r^2}$$

target

guide RNA

e^-

γ

e^+

H^2

The Science of
RICK
AND
MORTY

The Unofficial Guide to Earth's Stupidest Show

100

HO

$$t' = \frac{t}{\sqrt{1 - \frac{v^2}{c^2}}}$$

Matt Brady

$$\frac{kc^5}{R^2} + \frac{\Lambda}{3}$$

$$e^{i\varphi} = \cos\varphi + i\sin\varphi$$

m_1

ATRIA PAPERBACK
New York • London • Toronto • Sydney • New Delhi

$$\alpha = \frac{1}{4\pi\varepsilon_0} \frac{e^2}{\hbar c}$$

ATRIA
PAPERBACK

An Imprint of Simon & Schuster, Inc.
1230 Avenue of the Americas
New York, NY 10020

Copyright © 2019 by Matt Brady

Originally published in 2019 in Great Britain by Bonnier Books UK

First Atria Paperback edition October 2019

ATRIA PAPERBACK and colophon are trademarks of Simon & Schuster, Inc.

For information about special discounts for bulk purchases, please contact
Simon & Schuster Special Sales at 1-866-506-1949 or business@simonandschuster.com.

The Simon & Schuster Speakers Bureau can bring authors to your live event. For
more information or to book an event, contact the Simon & Schuster Speakers
Bureau at 1-866-248-3049 or visit our website at www.simonspeakers.com.

Manufactured in the United States of America

1   3   5   7   9   10   8   6   4   2

Library of Congress Cataloging-in-Publication Data is available.

ISBN 978-1-9821-2312-3
ISBN 978-1-9821-2313-0 (ebook)

For Mom & Dad—who always let me do my thing, even when it left them scratching their heads and wondering how I'd ever make a living at it. I think I have this figured out now . . .

For Aidan—who's going to show this part to his friends who don't believe that his dad wrote a science book about *Rick and Morty*. Yes, I really am that Aidan Brady's dad. Now pay up.

For Shari—for putting up with Writer Matt, Self-Doubting Matt, and Manic Matt for the last six months. And for being my rock.

And for Jack, for being a very good boy.

Contents

* ★ ★ ★ ★ *

Introduction

★ ★ ✦ ★ ★

"Break the cycle, Morty. Rise above. Focus on science."
Rick to Morty
Season 1, Episode 6: "Rick Potion #9"

You are going to learn things by reading this book, and someday, you might change the world because of it.

People as old as I am often hold *Star Trek* in a very special place. Weekly, the crew of the USS *Enterprise* (NCC-1701 and then 1701-D), the *Voyager*, or the space station *Deep Space Nine* amazed audiences with stories of a future where science could solve problems, equalize society, and create opportunities using technologies that we couldn't even dream of.

But looking back, what's a cell phone but Captain Kirk's communicator realized? The touchscreens that were ubiquitous on Captain Picard's *Enterprise* . . . how many times did you use a touchscreen today? Hundreds of other examples can be found throughout modern life. One could argue that *Star Trek* was the collective dream of future scientists and engineers that ended up creating the world they were shown.

Neither *Star Trek* nor *Rick and Morty* shy away from science. In episode after episode, it's there, in your face, an integral part of the world and something that's used by the characters. While *Star Trek* used science as a tool—a way to solve problems and to discover new things about the universe around us—*Rick and Morty* uses science as a toy: something to be tinkered with and used by Rick in the most unconventional and irreverent ways. But, in the end—science. Both shows get science into living rooms, and as a result, people think about science.

Rick and Morty is funny. It's irreverent, it's philosophical, it can be equal parts depressing and life-affirming, and it's also one of the smartest shows on television.

Think back to some of the biggest episodes with the most outlandish storylines—things like "Lawnmower Dog," "M. Night Shaym-Aliens!," "A Rickle in Time," or "Pickle Rick." They're all built around the newest, most innovative scientific concepts, and, more than that, they ask mind-blowing questions as a result:

* Can we make animals (or ourselves) smarter?
* Are we living in a computer simulation?
* Are there multiple timelines or realities, and can we move between them?
* Can a cockroach's brain be controlled with . . . chemicals?

Much of the world of *Rick and Morty* is built on cutting-edge science: hacking memories and dreams, the multiverse, cloning, biohacking and human augmentation, aliens, evolution, artificial intelligence, particle physics, and cosmology.

As a science teacher and writer, I'm always looking for ways to make science concepts more palatable for my students and my readers. To these ends, I like to think of pop culture as a Trojan horse. It's weird, interesting, and cool, and you let it settle into your brain, thinking it's just another bit of decoration or trivia that will get filed away. But then, late at night—or as you're watching/playing/listening to said pop culture—the secret door in the horse opens and the science comes spilling out into your walled city. Maybe it's a stray thought about making dogs smarter and giving them the ability to speak (and to question us about what we did to their testicles). Or maybe it's a question about multiverses that gets you wondering about where and how such things could exist. Or perhaps what gets out of the horse is a question about extracting and storing unpleasant memories.

And then, hopefully, you ask your science teacher about it, look it up online, watch a video about it, or even decide to read a book. If any one of these countless options happens, I'm happy. A seed of science has made its way into your brain.

Give that seed some water and some soil and see where it goes.

Maybe those scenes where someone grew or shrank got you thinking about and looking into the hows and whys of shrinking in our world, and maybe you learned a little something about it.

So is *Rick and Morty* an educational show? Sort of.

Is it inspirational? Hardly.

Does it offer a hopeful view of the future? Not at all.

But does it get its audience thinking and talking about science?

Since you're here, you've (hopefully) bought this book, or at least borrowed it from an extremely generous pal. So at the very least you're interested enough in *Rick and Morty* and science to be willing to crack this cover.

Rick and Morty may not turn you into a scientist, and you're probably not quite ready to write your first scientific journal article about dark matter after watching "M. Night Shaym-Aliens!" but it does get a large part of its audience thinking, wondering, and talking about science.

In this book, you're going to get a lot more of that science. The science behind dark matter, aliens, intelligence hacking, shrinking, growing, simulations, the multiverse, memory implants, cloning, and, of course, putting your consciousness into a pickle. All of that and more. And don't expect it to be any highfalutin science talk. When you close this book, you're going to understand what's going on and be ready to ask new questions. But don't worry: this isn't science class. *Rick and Morty* is irreverent—the least we can do in a book about the science of *Rick and Morty* is to be irreverent as well.

And in the end, after all of this, you know what I'd like to see? It's fifteen or twenty years in the future. There's a news conference with a scientist who's rumored to be fast-tracked for a Nobel because they've led a team that has conclusively proven we live in a multiverse. At that news conference—or later interviews, I'm not picky—they're asked how they got interested in the idea in the first place, and they say, "Did you ever hear of a show called *Rick and Morty* . . . ?"

CHAPTER 1

Alien Life

★ ★ ✦ ★ ★

The alien life on display in *Rick and Morty* is diverse, unique, and as weird as you can imagine on your funkiest, most messed-up day. From the adorably housefly-like Laarvians to the upsettingly waspy Gromflomites; from the vaguely anthropomorphized Birdman (ahem, Birdperson) to the uncomfortably anthropomorphized Squanchy, the nonhuman life-forms of *Rick and Morty* are comprehensive and disturbing. And they all come with evidence of complex social systems and are spread throughout the galaxy and multiple universes.

In our universe, though, we're not as lucky. We know of one planet where life has developed: Earth. So far, we haven't found alien life anywhere around us. But there's this itch, this feeling that there must be something out there, a desire to find interplanetary and intergalactic neighbors. Sure, this is fed in part by science fiction, but it also comes from a desire to not feel alone in the universe.

Let's keep things relatively simple, to start: life developed and exists on Earth. Our galaxy, the Milky Way, has between 100 billion and 400 billion stars. Recent exoplanet surveys have suggested that virtually all those stars have at least one planet, and some have several.

Just staying with the Milky Way, that's a lot of planets. It's been estimated that 10 percent of the stars in our galaxy are like our sun, and further exoplanet surveys suggest that one in five of those sun-like stars has an Earth-like planet orbiting it in its habitable zone—the region around a star where liquid water can exist on the planet's surface. If we say there are an average of 200 billion stars in the Milky Way, that gives us 20 billion sun-like stars—and from that, 4 billion Earth-like planets in their stars' habitable zones in the Milky Way alone. Expand the stars to include some non-sun-like stars that still may have habitable zones capable of supporting life, and that 4 billion can rise much higher—by some estimates up to 40 billion.

If you want to go to the extreme, there are around 100 billion galaxies in the observable universe, each with around 200 billion stars, 20 billion of which are like our sun, and, following the math above . . .

In short, our universe has billions and billions of Earth-like planets in it.

If we were playing the odds in our galaxy, we would assume that not every Earth-like planet has life on it. As a conservative estimate, we might say that one out of every thousand Earth-like planets has life on it, and one out of every thousand of those planets with life develops intelligent life. That's four to five thousand Earth-like planets with intelligent life on them in our galaxy alone. And that's not even counting planets and bodies that might be able to support life in our own solar system.

And yet, with all of the above weighing heavily in favor of life elsewhere, we haven't seen alien life anywhere. Earth is not a Galactic Federation outpost, and no one's neighbors have house parties with aliens.

What do we know about what—or who—might be out there? We'll start with the basics . . .

IS LIFE INEVITABLE?

At this point, Rick Sanchez has most likely created life dozens of times—some of it probably accidental, some of it to give his daughter playmates in Froopyland—and he most likely would scoff at the idea that creating life is something that's even challenging. And there are many scientists who would say he has a point.

As we continue to look at the universe, there's a growing feeling among some researchers that life may have a certain inevitable quality to it. The original, somewhat-based-in-religion view of life being something precious and extremely rare has been slowly changing over the course of the past half-century, as physicists, chemists, and biologists have been knocking on one another's doors more and more, asking their colleagues to think about some new thoughts and possible redefinitions of life. The thinking goes that if you have the right chemicals and the right amount of energy—all of which seem to be common—you'll eventually get life, of some sort. But this view was a long time coming and faced many pressures from different sides.

The common criticism against life arising spontaneously from "just chemicals" is that it goes against the Second Law of Thermodynamics, which, in short, says that disorder—the scientific word for this is "entropy"—always increases within a system. Life, on the other hand, increases order—you, as a person, have much more order than you as a pile of ingredients would. Order is created out of disorder. Matter is taken in and put into more and more complex structures by life. That's categorically the opposite of disorder.

But that's not the end of the story. Living organisms are entropy engines. We don't do much aside from collect energy and matter and break it down into simpler forms. Much of the chemical energy from the food we eat leaves our bodies as heat, which we basically see as waste energy. We can't do anything with it. That highly organized piece of plant or animal that was eaten isn't too organized afterward, either. Viewed this way, living organisms may affect the rate of entropy increase but don't cause an overall decrease. Life leaves disorder in its wake.

It's all good, Universe—your laws are still being followed.

While the view of the inevitability of life was discussed and even argued about throughout the early 1900s, the idea really came to life—pun almost intended—thanks to the work of Stanley Miller and Harold Urey in the early 1950s. The two scientists, first working at the University of Chicago and later at the University of California San Diego, re-created primordial Earth's atmosphere in a sealed, pressurized bottle. This remake of ancient Earth's "air" was historically accurate—methane, hydrogen, ammonia, and water, but no oxygen. When Miller and Urey sent electrical sparks through the gasses, simulating lightning, organic chemicals resulted, including more than twenty amino acids—the building blocks of life—and tholins, compounds that were present on the ancient Earth (and can, weirdly, also be found on Saturn's moon Titan). To be clear, no amoebae or worms crawled out of Miller and Urey's jar. No actual "life." Just the building blocks.

The thought leader for seeing life as an inevitable consequence of chemicals meeting energy is Dr. Jeremy England of the Massachusetts Institute of Technology. England's research suggests that as energy (for example, the energy from the sun) is added to groups of atoms (say, on the surface of the primordial Earth), they will naturally rearrange themselves to better absorb the energy, and then dissipate it as heat. Further, England explains, as the groups of atoms organize in a way to better absorb and release energy, there may be a tendency to self-replicate. This process is complicated and doesn't happen every single time energy hits a group of atoms, so don't expect a salt-based life-form to emerge from a pile of salt after shining a light on it for forty-five minutes.

But sometimes this outcome does happen, and as some scientists speculate, it continues on to its logical conclusion: life. England calls this idea "dissipation-driven adaptation."

If the start of life from non-life is so simple and so basic as far as physics and the universe goes, then it's hard to believe that it only happened once, just by sheer luck, and that it happened here and here alone. If you look at these ideas skeptically, though, there's a huge jump between energy helping atoms and molecules that could organize into structures that replicate themselves over and over and actual "life." The missing piece in this theory is the instructions. Even if physics, math, and probability can explain how you can get organic compounds and even some simple self-replicating structures, that's a far cry from life. As shown by the menagerie he created for Froopyland, Rick has clearly figured out how to go from chemicals to life, but we haven't. Perhaps there's some experiment out there that will show that life's "instructions" spontaneously generate as well, like the amino acids did in the Miller-Urey experiment.

That start of "life"—that's still out there, waiting to be found.

CAN I BORROW A CUP OF CARBON?

Most aliens in *Rick and Morty* probably have something in common.

Given how they all have expressed very humanlike qualities and needs, which would arise from a physiology like that of humans, it's not a huge leap to guess that the biochemistry of the majority of *Rick and Morty*'s alien life is based on carbon, just like all life on Earth. After all, many of the familiar alien species—Birdperson, Risotto Groupon, Squanchy, Plutonian, Arthricia, and the cat people of the Purge planet, and even Zeep Xanflorp inside Rick's Microverse Battery—are humanoid, and appear to be animals with similar or analogous structures to humans or other life on Earth. There are indications that many of the nonhuman life-forms of *Rick and Morty* are warm-blooded as well.

Saying there's a good chance that alien life would be carbon-based, like life on Earth, isn't species- or planet-specific bias. There are very good reasons why life on Earth is carbon-based: carbon is abundant, it has the ability to form bonds with up to four other elements, it can form stable double or triple bonds, and it can make chains, rings, and other structures. While we don't know for certain whether all alien life is carbon-based, the element may be the Occam's Razor of life. Wherein the philosophical principle of Occam's Razor says that the best explanation is the simplest, the biological analogue with carbon is that it's most likely the simplest foundation upon which to base life. Indeed, when it comes to Earth, carbon is a Swiss Army knife that life uses for a little bit of everything.

The remaining elements that we use for life (in order of their abundance in us) include hydrogen, nitrogen, oxygen, phosphorus, and sulfur. For decades, the thinking went that life could not exist without these other elements. But the thing is, we keep finding life that swaps out those elements for others and keeps on keeping on.

To better understand other elements that could be used for life, think back to the periodic table of your science classes. Elements in the same column are considered members of a group, or "family," and as such, they all share the same broad behaviors. It's why, for example, NASA's news about a microorganism that used arsenic rather than phosphorus was exciting but not wholly unexpected: arsenic can be found just below phosphorus on the periodic table.

Carbon's closest "relative" is the one located directly below it on the periodic table: silicon. Silicon is relatively abundant both on Earth and throughout the universe, but doesn't share a few of carbon's more important characteristics: it won't bond with

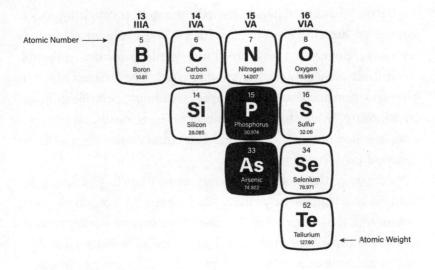

many types of atoms, and it has a limited number of shapes and forms it can take within molecules. As an option for alien life, silicon is attractive, which is why "silicon-based life" has been a staple in science fiction since the mid-twentieth century. The original *Star Trek* showcased the silicon-based Horta, and the xenomorph aliens of the Alien franchise are supposedly silicon-based, as are Korg from *Thor: Ragnarok* and the Kastrians and the Ogri from *Doctor Who*.

But the physics of completely silicon-based life remain a tough nut to crack, and would most likely require certain conditions that would be inhospitable to human life. Still, carbon and silicon could act together. Clays from early in the history of the Earth with silicon-containing compounds called silicates are thought to have been important in building structures upon which carbon compounds were able to attach. This idea has led some researchers to consider the possibility of organosilicon-based life.

Researchers at the California Institute of Technology were able to coax organisms to incorporate silicon into carbon-based molecules, even though this partnership does not occur naturally on Earth. The researchers achieved their goal through a process known as "directed evolution," which is essentially what humans have been doing for ages—breeding organisms for desired traits, while culling those with undesirable traits. The process of combining carbon and silicon together is exceedingly complicated, but by breeding generation after generation of microorganisms, scientists ended up with an enzyme that could create organosilicon compounds.

Admittedly, that's still a long way from Revolio Clockberg Jr., with his gears and boxy (perhaps silicon-based), transparent chest, but as with the discovery of the arsenic-incorporating organisms, science is constantly pushing the envelope of what we call life. And if it can be done in a lab, it's probably happening somewhere in the universe.

As for other candidates acting as a basis for life, some metals such as iron, aluminum, magnesium, tungsten, and titanium can form structures that are similar to those produced by living things, but life based on metal would have to exist in a very different environment and would stretch the idea of what we consider "life." Yet we've had those boundaries stretched before.

Carbon may be the easiest and most likely element for a basis for life, but *Rick and Morty* clearly plays with the idea that there's much more beyond carbon- and silicon-based life. Consider Fart (that's its name) from the episode "Mortynight Run"—a gaseous creature from another dimension that can change its atoms into any configuration as needed. That's a life-form that we're eons away from understanding, and one that, if ever discovered, would stretch our definition of "life" in new directions.

While the idea of a fart as a life-form is laughable, let's remember what it said to Morty: "Carbon-based life is a threat to all higher life. To us, you are what you would call a disease. Wherever we discover you, we cure it." It is interesting to see that, just as life on Earth may carry a carbon bias, or be "carbon chauvinists" as the astronomer Carl Sagan once described himself in regards to alien life-forms, life based on some other collection of elements like Fart can have its own biases as well. Perhaps Fart's people and other life-forms don't create as much disorder as highly inefficient carbon-based life-forms. Maybe we remind them of an ugly past of their universe that they want to forget. Elemental bias can cut both ways.

In a larger context, Fart's statement to Morty was chilling, and a reminder that we might not be all that when it comes to life in the universe. It also led to one of the sharpest points of moral ambiguity in the series to date: Fart was being held captive and marked for assassination by Krombopulos Michael with the Rick-supplied antimatter weapon. If you were a carbon-based life form, you might not want Fart wandering around either.

WILL WE MEET GOO OR GABLOVIANS?

As has been depicted over and over on the series, the aliens of *Rick and Morty* take many forms. On the humanoid side, you've got the insectoid Gromflomites of the Galactic Federation, the cat people of the Purge planet, the Bird People, the Garblovians (of the "argle-bargle" language), the telepathic Krootabulans, the machine-based Gear People, the Zigerions, the Meeseeks, and dozens more, named and unnamed. And that's just the humanoid side, without getting into all the different species that showed up in "Ricksy Business."

Throughout their travels, Rick and Morty have encountered alien life-forms that appear to fill specific evolutionary niches in our familiar tree of life, such as tentacled cephalopod types, various parasites, annelids (eye worms), crustaceans (crab spiders), and reptiles (Crocubot).

But the level of complex life shown in the jungle in "The Whirly Dirly Conspiracy"—the worm that ate Jerry and the larger animal

Rick saddled and rode with Jerry underneath (the one whose ball sack kept slapping Jerry in the face)—is probably quite rare in our universe.

To get to that level of complexity, there would need to be a web of simpler life underneath those species—the bacteria, the fungi, the molds, and the life of the respective home worlds of the various alien species. That's most likely what we're going to find when we do start finding life on other planets, asteroids, and moons.

In our experience, complex life takes time. A lot of time. Let's run it down: our universe began with a Big Bang around 13.7 billion years ago. Lots of stuff happened after that, but about 4.5 billion years ago, the Earth was formed. Life started on Earth with single-celled prokaryotic organisms around 3.8 billion years ago, which is pretty early on in Earth's history, relatively speaking. Prokaryotes are "simpler" cells than what we have in our bodies. They're cells without a nucleus or any membrane-bound organelles inside them. Their DNA is either free-floating or found in the form of a circular plasmid. Prokaryotes are still around today, most commonly showing up as bacteria.

It took about another billion years before things got a little more complex—eukaryotes, cells with their DNA wound up and tucked inside nuclei and organelles bound inside protective membranes, showed up around 2.7 billion years ago. Sexual reproduction first appeared 1.1 billion years ago, and then things really got cooking. Evolution started in earnest.

Multicellular eukaryotes evolved a few times, but by 600 million years ago (MYA) they succeeded, both in the oceans and on land. Arthropods (today's insects, arachnids, and crustaceans) showed up 570 MYA, fish arrived 530 MYA, land plants 475 MYA, amphibians 370 MYA, reptiles 320 MYA, dinosaurs 225 MYA, mammals 200 MYA, great apes 14 MYA, members of the genus Homo 2.5 MYA, and Homo sapiens, our species . . . just 200,000 years ago.

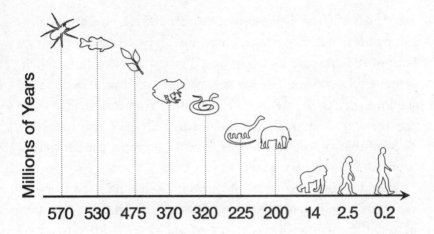

If you ever start feeling full of yourself, just remember that last number in relation to the formation of the Earth. Humans have been around for 0.004 percent of the Earth's history.

And, of course, the dominant form of life didn't have to be human at all. The fact that humans are the most intelligent creatures on Earth is due to a very specific combination of evolutionary pressures that include everything from climate to Earth's position in the solar system to the size and shape of our orbit around the sun and fortuitously timed large-scale extinction events. All of those things, taken together, are pretty indistinguishable from luck.

For example, the asteroid that smacked into Earth in the Yucatán Peninsula 66 million years ago, which ended the age of the dinosaurs, wasn't just bad news for the thunder lizards: it wiped out about 75 percent of all life on the planet. The debris lofted into the atmosphere by the impact, perhaps aided by truly apocalyptic volcanism triggered by the collision, caused a climate change that dropped the planet's temperature an average of 14 to 18 degrees Fahrenheit and left behind a crater called Chicxulub, which can be found under the water of the Gulf of Mexico to this day.

More than just a game of idle speculation, recent research has asked, what if the asteroid had hit somewhere else on Earth—somewhere that didn't have limestone, gypsum, and an oil-rich layer of sedimentary rock that would throw up dirty, carbon-rich, planet-choking soot, somewhere where the water wasn't so shallow—say, in the Atlantic Ocean rather than the Yucatán? The argument is easily made that mammals rule the Earth today because of that asteroid and the sun-blocking, temperature-lowering soot that it threw up into the atmosphere.

Given the Earth's rotation, if the asteroid had hit a few minutes earlier or later, it may have hit somewhere else on the planet, perhaps in the Atlantic Ocean, which would have absorbed much of the energy. If that had happened, we'd most likely never know about it. Maybe reptilian scientists, descended from dinosaurs—rather than the early mammals that were running all over the Earth just half a million years after the impact—would be using their own submersibles to dive to the bottom of the Atlantic (or whatever they would call the body of water) to see what's down there and finding an amazing surprise. On that alternate, reptilian-dominated Earth, perhaps audiences would read about this amazing discovery that almost ended all reptilian life on Earth 66 million years ago on their version of the Internet.

We don't even have to speculate if we want to understand how lucky breaks made us what we are. Earth's orbit has subtly changed throughout its history, causing climate changes ranging from mild to severe. Africa's Rift Valley, which has been theorized to be the birthplace of modern humans, has a peculiar ability to magnify these climate changes, and as a result, beginning over 2 million years ago, the area went through cycles of feast and famine. These cycles forced our ancestors to evolve into a new species with bigger brains in order to survive.

So, yeah—we're lucky.

Looking at the aliens of *Rick and Morty* through this lens, we've got no reason to assume that these (admittedly fictional) species don't have their own harrowing story of near-extinctions and lucky breaks that allowed them to come out on top so that one day they could go to a party at Rick's daughter's house. But again, complex life takes lots of time and luck. If you don't have both, you're not going to the party.

OLDER THAN US?

In the universe of *Rick and Morty*, there are dozens and dozens of alien species that are far more advanced technologically than humans, and as best we understand it, the development of technology takes time. So, actually, *Rick and Morty*'s aliens are most likely older than humans on Earth.

Let's take ourselves as the example again: the Earth is about 4.5 billion years old, and the oldest fossils date back 3.5 billion years for ocean-dwelling organisms and 3.22 billion years for those that lived on land. As mentioned earlier, it took about 2 billion years for single-celled organisms to evolve into multicellular organisms, and another billion years to get to a point where we'd even recognize them as organisms. With all this talk about age, don't be fooled into thinking our solar system is anything special, or that it was anywhere near "first." Our sun was not one of the universe's first stars. We're a little late in the universe's timeline. It's very possible that the first life-supporting planets have been around since the universe's early days, giving these otherworldly beings a potential billions-of-years head start on us.

The universe is around 13.7 billion years old. At a conservative guess, the first galaxies formed about one billion years after the Big Bang. Our solar system took about 5 billion years to go from a gaseous cloud around a sun to distinct, mature planets, so let's use that as an average time for solar system formation. It took (again, in our experience) 2 billion years to go from a single cell to multicellular life, and then another billion years for that early multicellular life to evolve into something that we would recognize as creatures like what we see around us today. All told, that's about 9 billion years from nothing to life. So, in theory, and using our measurements, the first life in the universe could have shown up . . . 13.7 billion − 9 billion = 4.7 billion years ago.

Of course, it is possible that life could have started before that 9-billion-year line. Again, using Earth life as a meter stick, it took a long time to move from prokaryotic life without a nucleus in the cell to eukaryotic life with DNA safely tucked away inside a nucleus. This huge milestone for life took a billion years to happen on Earth, but on some alien world, it could have happened much sooner, or later—there could be a planet where alien prokaryotes have been bumping around for five billion years and haven't been able to get eukaryotic just yet.

But if we take our measurement based on our Earth-bound experience, the 4.7 billion years that an alien race could have gotten started before us feels incomprehensible. There could have been civilizations that lived, thrived, and died out, the cycle repeating itself in different systems in different galaxies over and over and over before our sun sparked into fusion-driven life.

Besides rocks, there's nothing on Earth that we can point to and correctly use the word "billions" when it comes to age. "In a billion years, humans . . ." doesn't even make sense. A billion years ago, we humans weren't here, and a billion years from now, if we don't kill ourselves, we probably won't be recognizable as "humans." Who knows—maybe in a billion years, we'll have turned from primitive pursuits like art and science in order to develop the technology to move planets and feed on the talent and showmanship of less-evolved life-forms, like the Cromulons do in *Planet Music*.

But given that the clear majority of the aliens in *Rick and Morty* share similar social structures, morals, larger belief systems, and other elements of culture and society with humans, even with their advanced technology, we're probably less than 100,000 years behind any of them. This number comes from looking at human society through the ages: the basic units of human organizational structure, society, culture, and belief systems have remained relatively unchanged since we settled down and started farming between 12,000 and 20,000 years ago. Sure, technology has changed and advanced in that time, but has humanity? It's hard for us to say how long a total change of a society takes, because we've never really seen one happen.

One could argue that the Internet has allowed a connectedness like we've never seen, and we may be poised on the edge of a great leap forward, but that's a solid maybe. An easier argument is that we're still as tribal, greedy, and prone to violence as we've always been. Just like the aliens shown in *Rick and Morty*, even with their advanced technology.

The fact that *Rick and Morty*'s alien life has many negative behaviors that are so recognizable to us, like greed, a thirst for power and domination at the expense of those who are weaker, and massive consumption of resources, is why alien species are used more as metaphor rather than scientifically accurate examples of life. They provide us a chance for reflection. What would we be like if we were coolly rational and logical all the time? Looking at aliens in science fiction is a way for us to examine ourselves.

BACK TO THE CARBON-BASED, SIMPLER LIFE

Despite the appearance of the Plutonians in the episode "Something Ricked This Way Comes," we're not going to find any alien civilizations in our own solar system.

Not that that's stopped people from trying. Theories about life in our solar system have existed through history, and have often corresponded to our understanding of life (and science). Take Mars, for example: consolidating what was known about the planet by the mid-nineteenth century, scientist William Whewell suggested in 1854 that Mars had land and seas, and possibly life. In 1877, the Italian astronomer Giovanni Schiaparelli observed lines on Mars that he called *canali*, meaning "channels." The word was mistranslated into "canals," which fed the idea that Mars had an intelligent civilization that built canals to transport water between regions of the planet. Building upon this, American astronomer Percival Lowell published the book *Mars and Its Canals* in 1906, followed by *Mars as the Abode of Life* in 1908.

While other scientists dismissed Lowell's ideas—namely because the "canals" were optical illusions—the public held fast to the idea that there were creatures of some sort on Mars. Though Lowell's ideas fueled science fiction throughout the twentieth century, scientists were not fans, and pointed to the growing evidence that showed Mars to be a dry, sterile world with harsh conditions and no life.

This view that Earth is special and the rest of the solar system is inhospitable held for decades—not just for Mars but also for the greater solar system, and perhaps farther out as well. Life was seen as vanishingly rare, requiring a very precise and complex set of conditions that had to be matched exactly.

But that view started to weaken as creatures called extremophiles made their presence on Earth known.

The study of extremophiles—organisms that make their homes in physically or geochemically extreme locations previously thought to be unable to support life—got a boost in 1977 with the discovery of life near volcanic vents on the ocean floor. Encased in eternal darkness and surrounded by water that reaches over 550 degrees Celsius (the pressure prevents the water from boiling), life endures. Two- to three-meter-tall giant tube worms call these places home, along with clams, one-eyed shrimp, blind crabs, and Pompeii worms that live in water reaching nearly 100 degrees Celsius in temperature. Unlike surface-dwelling life, which gets its energy from photosynthesis, as plants capture the sun's energy, this ecosystem on the ocean floor thrives on bacteria and microorganisms that capture energy from the heat and chemicals coming out of the vents via a process called chemosynthesis.

The environment near the hydrothermal vents is thought to be like that of Earth of billions of years ago, and as such, many biologists suggest that the life found near hydrothermal vents may be very similar to the first life on Earth. All surface life, the suggestion goes, evolved from the organisms that first appeared near primordial Earth's analogues of hydrothermal vents, rather than on the surface.

As mentioned earlier, in 2010, NASA announced it had found a microorganism that used arsenic instead of phosphorus in its DNA. The organism was found in California's Mono Lake. The lake is an extremely salty, alkaline environment that could easily be used as a background for any television series set on an inhospitable alien world. Other scientists had issue with the original findings and claims that the organism was "arsenic-based life," but follow-up studies did confirm that the organism was able to live in an environment which had very high levels of arsenic and only minute levels of phosphorus. While it may not have incorporated arsenic into its DNA, it was another example of life in extreme conditions.

This seems like a good place to point out that Rick insulted the pawnshop owner at the start of "Raising Gazorpazorp" by telling him that his species eats sulfur. In our world, sulfur-eating species are bacteria that live near hydrothermal vents, getting their energy from the oxidation of sulfur. Rick was clearly using it as an insult about how advanced the store owner's species was.

Organisms living near hydrothermal features and those that live in arsenic-rich environments are just two of a growing collection that includes, among others:

* Endoliths: organisms that live inside rock.
* Acidophiles: organisms that can endure highly acidic conditions.

* Piezophiles: organisms that can live at high pressures.
* Cryophiles: organisms that can live at extremely low temperatures.
* Radioresistants: organisms that can live in the presence of harmful radiation.

And that's not counting the list of extreme locations where organisms have been found to survive, such as high in the atmosphere, in a high-gravity test chamber, scavenging gasses from the air to stay alive in Antarctica, or in liquid asphalt lakes, like Pitch Lake in Trinidad and the famous La Brea Tar Pits in Los Angeles.

Extremophiles changed and are continuing to change our definition of life, and with it, our ideas about the conditions under which we may find life on Earth and elsewhere.

FINDING ALL THIS POSSIBLE LIFE: WHERE?

So, where can we begin looking for life? Let's start locally.

We have looked at every planet in our solar system, as well as dozens of moons and asteroids, since we could reliably get probes off Earth, starting in the late 1950s and early 1960s as the United States and the Soviet Union competed for space bragging rights. Not every mission was looking for life, but there was always a hope that if something larger than a small bug lived on the probe's target, it would just happen to move in front of the camera or other sensor.

The lack of signals from life-forms on planets such as Venus and Mars backed up the idea that conditions there were most likely inhospitable to intelligent life. Only recently, as probes and sensors have gotten better, have we been able to determine the chemical makeup of the soils and rocks of Mars, for example. Those instruments have shown the presence of organic molecules such as methane, amines, and others in many locations throughout the solar system. Just to clear up some hazy memories from chemistry class: while they are referred to as "the building blocks of life," organic molecules mean those produced by or necessary for life, not just all compounds that contain carbon. Plastics contain carbon. Lots of carbon. But no one would really argue that plastic is produced by or needed for life.

Tests by NASA's *Curiosity* rover, which landed on Mars in 2012, have shown organic compounds present in rocks formed at the bottom of an ancient lake bed in the Gale Crater. If we wind the clock back about 3.8 billion years, just when life was developing on Earth, Mars had very similar conditions as we had here.

Given these findings, it's tempting to start making a speculative claim for life on Mars, or to tweak that line of reasoning and imagine that life developed on Mars before it appeared on Earth. An asteroid hit Mars, and part of the asteroid was blown back into space—with some hardy Martian microbial life attached to it—only to impact on Earth, with the microbial life intact. A few million years' worth of evolution later, and maybe . . . we're all Martians.

Elsewhere in the solar system, certain types of organic compounds were found on the surface of the dwarf planet Ceres, which is the largest object in the asteroid belt between Mars and Jupiter. Spotted by the *Dawn* space probe, these organic compounds were "localized" near a crater named Ernutet. That is, they were concentrated in that one area rather than evenly distributed across the surface, which is what you would expect to find if the compounds were always there or arrived via a collision with something carrying them. Moreover, the types of compounds were not what was expected—they were long carbon chains rather than rings of carbon atoms, which are far sturdier than chains. All in all, the evidence somewhat resembles what we find around hydrothermal features on Earth. Hot water, organic molecules, and a clay-based substrate to work with, which is like what biologists say might have been the original conditions for life on Earth.

Looking at two "ocean moons" of our solar system, the *Cassini* spacecraft flew through an organic-compound-laden geyser of frozen water that regularly sprays into space from Saturn's moon Enceladus, while organic compounds apparently line the cracks of the surface ice of Jupiter's moon Europa, suggesting that they come from the waters below. Farther out, organic compounds called polycyclic aromatic hydrocarbons (PAHs)—ring-shaped carbon molecules needed by all forms of life—have been spotted in a galaxy 12 million light-years away. And further studies indicate that stars create and eject complex organic matter into interstellar space. Wherever you look, you'll find it, apparently.

But, as skeptics will point out—and they are correct—the presence of organic compounds does not mean life. It just means that organic compounds are present. Organic compounds can be formed by geologic processes or brought in by meteorites on which they may or may not have been formed by life. So, the presence of organic molecules does not mean the presence of life, but . . .

A common analogy for hunting for organic molecules on Mars is hunting for gold. If you wanted to find gold, you wouldn't go out and just start digging haphazardly in your backyard. You'd do the research, look at the maps, and find the places where you'd expect gold to be found—where the gold should be, because all the other conditions are right. That's what hunting for organic molecules has been like on Mars. *Curiosity* isn't just digging willy-nilly in any location it happens to roll by. It was specifically directed to the Gale Crater, a dry lake bed that possibly, long ago, had living organisms in its waters. But again, having the building blocks of life does not necessitate life. Their discovery is not a smoking gun but rather a tantalizing clue.

It's easy to get swept up in the excitement of the discoveries, but it's frustrating when no life is found. Many researchers believe that while we haven't found any life out there, it's just a matter of time.

ALL THIS LIFE—HOW DO WE FIND IT?

If we choose to believe that there is life out there, the next question becomes: How do we find it?

First things first—we're relatively new to planetary exploration, so we can cut ourselves a little slack about not having found any alien life yet. As mentioned earlier, in the long view, humans have only recently started looking for life on other planets. Looking for extraterrestrial life wasn't that big a deal early on in space exploration, so we weren't carrying the correct sensors and tools.

Even now, the hunt for life in the solar system moves at an agonizingly slow pace, and in some ways it's locked in time. Take the *Curiosity* rover on Mars. Part of its mission is to determine if life ever arose on that planet, and it was equipped with modern tools to achieve that goal. However, *Curiosity* was the winning proposal for a rover mission in 2004, with a proposed launch of 2009. Delays pushed the launch back to November 2011, and it has been on-site since August 6, 2012.

Curiosity's tools are great for the job, and with sophisticated NASA engineering and an Earth-bound engineering team still working on the mission, *Curiosity* (like the rovers that came before it) has already outlived its initially stated life span of two years, and will continue to work on Mars indefinitely.

Of course, while it will still work, it will work with dated technology. Our space exploration missions carry technology that was cutting-edge at the time of lift-off. Think about *Curiosity* on Mars. As big as an SUV, and rolling around with technology that was trailblazing when the iPhone 4 was the most advanced cell phone you could get.

And the mass of your tools is a consideration. Each kilogram put into space costs tens of thousands of dollars and requires rockets and payloads to be reengineered. Space missions only carry tools that will pay off. *Curiosity*, as much teeth gnashing as it may cause those of us on Earth, does not carry a microscope capable of spotting microscopic fossils of bacteria. Assistant director for science communication at NASA's Goddard Space Flight Center Dr. Michelle Thaller once jokingly complained that there was no backward-mounted camera on *Cassini* capable of showing the spacecraft itself. Since *Cassini* flew through a saltwater geyser from Saturn's moon Enceladus, Thaller said, there could be Saturnian analogues of brine shrimp or other organisms smashed on the craft like bugs on a windshield.

So on-site and flyby missions are doing the best or sometimes better than the best they can with what they've got. More Mars missions are planned, while trips to Enceladus and Europa are on the drawing board. But that's just our solar system.

If we want to find life elsewhere in the galaxy or the universe, we need to look at exoplanets: planets orbiting stars other than the sun.

In terms of science history, exoplanet science is brand-new. While the idea of planets orbiting other stars has been around since the sixteenth century, when Italian philosopher Giordano Bruno suggested that, following the Copernican view of Earth, other stars would have their own planets, the science really took off in 1992 with the first detection of two planets orbiting a pulsar. As of this writing, there are 3,797 confirmed exoplanets orbiting stars both near and impossibly far away, and as those planets move in front of their stars, signs of life can be spotted.

Sensors look at the thin ribbon of light from the star that travels through the exoplanet's atmosphere, break that light into its components using spectroscopy, and, from those pieces, determine the gasses that are in the atmosphere of the planet. Life produces atmospheric biosignatures, so the presence of oxygen in the atmosphere, for example, is a dead giveaway that something on the planet is producing it—otherwise, chemical processes would have locked it away in other compounds. Water vapor, methane, and a few other chemicals are also strong indicators of life, especially if they show seasonal variation. Other research suggests that if we're looking for intelligent life, we should look for industrial pollution among the atmospheric gasses. The strength of both approaches is that Earth could be used as a model for comparison. Although if extraterrestrial life is not based on the life chemistry that we're used to, all bets could be off, and the exoplanets' atmospheres could have life signs that are new gas combinations that are needed for life—but not as we know it.

One final way of detecting alien life is aimed at looking for intelligent life that uses technology.

Modern programs with the label "Search for Extraterrestrial Intelligence" (SETI) were begun in 1960 at Cornell by Frank Drake. All the various programs were established with the mission to explore, understand, and explain the origin of life in the universe and the evolution of intelligence. That's a mouthful, but in terms of what we're talking about here, SETI scans the skies with a variety of telescopes, looking for any signals of intelligent extraterrestrial life, or "technosignatures."

For many years, SETI research looked and listened and found almost nothing. The "almost" came in 1977, when a strong signal at a specific wavelength was detected by the Big Ear radio telescope at Ohio State University. The signal, which came from the direction of the constellation Sagittarius, lasted for seventy-two seconds and had a frequency of 1,420 megahertz. That frequency is important—it's the same frequency that's naturally emitted by hydrogen atoms and theorized to be a kind of universal "ID card" for technologically advanced civilizations. The observer who spotted the signal was so impressed by it that he circled the specifics on the printout and wrote "WOW!" to the side, inadvertently giving it the name the "WOW!" signal.

Emitting a signal at 1,420 megahertz, the thought goes, shows that you're clever. It's like a little bit of intellectual bragging.

But it never happened again. In the forty-plus years since the signal was first detected, there has been complete radio silence. And researchers have been looking, with better and more sensitive instruments. In 2012, we sent a message back to the area of the signal's origin—although that was largely a publicity stunt and consisted of 10,000 Twitter messages with the hashtag #ChasingUFOs. #facepalm.

But SETI programs keep developing new and different methods of looking for intelligent life. But so far . . . nothing.

INTELLIGENT LIFE: WHERE IS EVERYBODY?

Let's run back through our journey:

* There are billions of Earth-like planets in our galaxy.
* Based on physics, life can be seen as an inevitable consequence of having the right stuff in the right place at the right time.
* Carbon-based life is probably the easiest on-ramp to the road to life.
* Life can exist in virtually any environment—from gentle to harsh—on Earth.
* Organic molecules that are consistent with the presence of life can be found virtually everywhere we look.
* Given enough time and luck, simple life can evolve into complex life-forms.
* At least in our sample (Earth), one of those complex life-forms became technologically advanced and is reaching out from its home planet.

In short, given the vast number of locations that could support life in the universe and the probability that some of that life would be intelligent and technologically advanced, the lack of evidence of extraterrestrial life is stunning.

So where is everybody?

This was the question that Nobel laureate Enrico Fermi asked his colleagues one day in 1950 as they walked to lunch, during a discussion about alien life visiting Earth. There was laughter and joking about the question, but this was Fermi, the physicist thought of by many as the architect of the nuclear age. If you don't already know his name, you might want to check out element number 100 on the periodic table—not everyone has an element named after them. When Fermi asks a question such as this, it's not just lunchtime jibber-jabber to him. The great physicist was famous for his almost supernatural talent for estimation and calculation. He would pose difficult questions, such as how many piano tuners are in Chicago. If you followed his rationale, you could get an answer that was very close to, if not exactly, the actual answer.

Fermi's obsession with calculating answers to barely under-stood questions is a recognizable characteristic of great minds, and one shared with Rick Sanchez when he has to calculate how to construct the microverse in his ship's battery or figure his way out of the time fractures of "A Rickle in Time."

Fermi ran some numbers and concluded that, if there was in-telligent, extraterrestrial life in the universe, we should have been visited—many times. This is what's called the Fermi Paradox: when estimation suggests something that isn't true. In this case, we should have been visited by alien life, but we haven't been.

The Fermi Paradox has been rigorously debated and expanded upon since its initial utterance. Perhaps the most famous explo-ration down similar lines of reasoning came from the previously mentioned American astronomer Frank Drake in 1961, and has become known as the Drake Equation.

Drake sought a way to look at the probability of intelligent life in the universe, and, as such, developed an equation that allowed a calculative approach to the same question Fermi asked. The equation wasn't meant to be formally solved, but to serve as a touchstone for further study. Drake came up with seven factors that he felt were crucial to predicting the number of detectable civilizations in the galaxy.

The equation has been added to and modified by others over the years, but the basis remains the same:

$$N = R^* \times fp \times ne \times fl \times fi \times fc \times L$$

* N = the number of detectable alien civilizations
* R^* = the average rate of formation of suitable stars
* fp = the fraction of stars that have planets
* ne = the average number of habitable planets per star
* fl = the fraction of habitable planets where life begins
* fi = the fraction of planets where intelligent life evolves
* fc = the fraction of planets where intelligent life is capable of interstellar communication
* L = years a civilization sends signals into space

When Drake first formulated the equation in 1961, many of the variables were estimates that no one had any way of knowing. As we've learned more about the abundance of exoplanets and other variables in the equation, some values have come into sharper focus, but again, the equation isn't meant to be a solvable problem but rather one that serves as a basis for further discussion and study.

While Fermi's Paradox may suggest that a vast number of alien species are making a point not to visit us, Drake's Equation tries to attach a probability to the existence of alien civilizations that we can talk to. Even plugging extremely conservative numbers into Drake's Equation, both ideas suggest that if we're looking for intelligent neighbors in the universe, something or someone should be out there.

So Rick throws a party, and Beth and Jerry's house is full of aliens. Birdperson and Tammy's wedding—loads of aliens. But we have none. We're not even at the level of ET.

Not that we should use *Rick and Morty* as a guide for our universe, but why aren't we hearing from or spotting aliens?

One of the major reasons might be that we're stuck in the present. We don't know about the past or the future, and, worse yet, being the winners of life's game of evolution has made us horribly biased. We think we know how we got here, but we only really know the larger process as a rough sketch. In the same way, we don't know what the future holds.

This is called the Great Filter Theory, the belief that there could be some filter somewhere along our progression that acts as a bottleneck—one that we passed through, but since it was so long ago, we don't know about it; or, because it's coming up, one that we can't know about.

Think about the move from prokaryotes to eukaryotes. That was a huge step in the journey of life. It wasn't easy, it wasn't happening a lot, and it took a billion years. That step may not be as common as we think. We may only be here because one cell on one really good day made the change, and maybe in terms of the universe, it happens only once every five billion years.

The same could be argued about moving from single-celled organisms to multicellular organisms, or the filter could be the random chance that atoms collect into molecules that replicate. Turn any one of those passageways to the future into an insurmountable wall, and all forward motion stops, and you never get complex life.

Or the filter could lie in our future. Life could develop complexity and technology, but only in one in a billion tries does life manage to get off its home world. We often seem to be moving toward a bleak future that seems to suggest technologically based civilizations poison their planets or kill themselves due to primitive urges before they can move off-world or kill themselves with their advanced weapons.

If the Great Filter Theory isn't depressing enough, coming up with other "solutions" to the Fermi Paradox has become a cottage industry, and the answers range from silly to ultra-depressing and frightening. For example:

The Interstellar Zoo: We are a "kept civilization," pets of a more advanced civilization, and the "keepers" are the aliens that occasionally come by for a visit or a quick anal probe but otherwise leave us alone.

We're It: There's lots of life out there, and some is even intelligent, but we're the most intelligent of all.

Scale: Space is really, really big, after all, and no one has figured or will figure out how to travel faster than light and contact other species. Our "houses" are just too far apart.

Physical Limitations: Intelligent life exists on super-Earths that have a higher gravity than we do, and as such, rocket launches are nearly impossible, or that life may have developed too close to a black hole to make space travel possible.

The Cosmic Graveyard: Intelligent life so rarely pops up that when it does, it's separated by millions of years, so that those other civilizations are long dead, either by their own hand or age.

The Dark Forest: Every civilization needs matter and resources, but those are finite; therefore, every civilization other than yours is a threat—so you shut up, hide, and don't make any noise that would attract predators as you hunt your own prey.

And, of course, the possibility exists that we just can't recognize alien life because it's so different from our own or so advanced that it has no need to interact with us. After all, ants don't know humans as individuals, nor do they know exactly what we are in relation to them. To ants, we're a force of nature—something akin to a storm or a rock rolling down a hill that crushes your buddies but leaves you standing.

As the science fiction author Arthur C. Clarke said, "Two possibilities exist: either we are alone in the universe, or we are not. Both are equally terrifying."

GETTING LIFE FROM HERE TO THERE

In *Rick and Morty*, spreading life around the universe is as easy as forgetting to wipe your feet after a particularly gross adventure, or when a space-bug smashes on your windshield like at the start of "Look Who's Purging Now." But in our reality, without faster-than-light travel, a simpler idea has been proposed to move life from place to place: panspermia.

Suggested by Fred Hoyle and Chandra Wickramasinghe in 1974, the idea is that complex organic molecules and nucleotides could move between planets on interstellar dust, comets, or asteroids. The organic, molecule-laden debris would, in a sense, "infect" other planets the way one sneezing toddler can infect a whole train car of passengers. Given enough time, the organic molecules could combine and ultimately begin life. To their credit, Hoyle and Wickramasinghe made this suggestion before we knew that organic molecules can be found just about everywhere you look in space.

The idea has been further developed to suggest that organisms could even hitch rides between planets where life could be supported. But those organisms would have to be extremely tough. Open space is swimming with hard radiation from nearby stars and other sources. Ultraviolet radiation, X-rays, gamma rays, and charged particles can cause strand breaks in DNA, which could kill the organism outright, or induce so many mutations in its genetic material as to make it unable to reproduce, or not able to live in the environment. To live in space unprotected, an organism would have to have some incredibly hardy means of protection.

Research on the survival of organisms in open space is mixed: in 1966, bacteriophage T1—a type of virus that is abundantly common on Earth and infects bacteria—as well as *Penicillium roqueforti*—the fungus responsible for blue cheese—were exposed to the hard vacuum of space as part of the Gemini missions. Both organisms became inactive due to UV radiation after a short period of time. In the 1990s, the European Space Agency (ESA) tested spores of various organisms in space, both naked to space and inside artificial meteorites. All suffered enough DNA damage to kill the organisms, although it was shown that the spores' survival rates increased if they were encased in glucose, a simple sugar used by living organisms for energy. The ESA also showed that some bacteria can remain viable in open space if shaded from direct UV radiation. Lettuce seeds and some lichens can also manage as well.

As for organisms larger than what you would sneeze out, the ESA's Foton-M3 mission took 3,000 tardigrades (also known as water bears) aloft see if they could survive exposure to open space. Tardigrades are the tiny badasses of the animal world. At only half a millimeter long, various tardigrade species have been shown to be able to survive in temperatures from minus 458 degrees to 300 degrees Fahrenheit, being dessicated, and being frozen for thirty years, waking up to lay viable eggs and eat an algae snack. The ESA's mission added "open space" to the list of things that don't really bother tardigrades. Cosmic radiation, freezing temperatures, zero air pressure, and no water—make life as harsh as you can, water bears won't care.

Research continues to see how long tardigrades can survive in open space, as well as whether they can survive a meteorite entry through an Earth-like atmosphere if they were buried deep enough inside the rock. But just so we're clear: although you can draw conclusions about organic compounds, some bacteria, and even lichens and water bears surviving in open space, no one is saying that they did, or that they played a key role in the origin of life on Earth. It's all just adding to our interesting picture of things to consider.

Since 1996 there has been a fierce debate about whether fossilized bacteria was found on a meteorite from Mars in Antarctica. Supporters claim that the meteorite was most likely a life-carrying chunk of Mars that was knocked off the Red Planet by a powerful impact that, thousands of years later, made its way to Earth. While skeptics point out that the features cited by scientists as evidence of bacterial formation could have had other origins, many have trouble explaining the presence of microscopic magnetite crystals that are produced by microbes, not geology. Shock waves, like those experienced by the meteorite as it came through Earth's atmosphere, have been known to create crystals of the same type as those in the meteor, so the final judgment still isn't clear. But that's the way good science works. The initial finding was greeted by strong skepticism, which resulted in rigorous testing.

Earth, too, has its own connection to possible panspermia. Like Mars, ancient Earth could have had life-bearing chunks knocked off from asteroid impacts (the asteroid that wiped out the dinosaurs was probably strong enough to knock stuff out into space), meaning that we could be the source of the "infection." There's a chance that any life we find elsewhere in our solar system originally came from Earth.

But ancient rocks and ejecta from collisions aren't the only things that leave our planet.

In September 2017, the *Cassini* spacecraft ended its thirteen-year mission observing Saturn with a fiery dive through the ringed giant's atmosphere. Like a dive into Jupiter performed by the *Galileo* probe in 2003 and a similar one that will be performed by *Juno* into Jupiter in 2021, the maneuver was designed to completely incinerate the spacecraft.

Torching our spaceships isn't about avoiding a V'Ger scenario or aliens getting their hands on our tech. With *Cassini*, the case was made plain by NASA—the probe had sent back copious data suggesting that Saturn's moons Titan and Enceladus could possibly support life. Burning up *Cassini* was the only way to make 100 percent sure that no extra-hardy microbe stowaways could possibly contaminate the moons with life that originated on Earth.

And, finally, NASA famously opened a public application process for a "planetary protection office" in 2017. While the jokes came fast and furious, NASA was serious. It was looking for a qualified applicant who understood the biochemical and medical risks inherent in humans possibly contaminating other planets, as well as those of alien matter infecting Earth. The position was created as part of the Outer Space Treaty of 1967, which stipulated that exploring signatories take steps not to infect the destinations, as well as to protect Earth from being infected from space. Nations agreeing to the treaty had to guarantee that any mission would have less than a 1 in 10,000 chance of contaminating an alien world.

Panspermia may have its skeptics, but a lot of people on Earth are taking steps to make sure a version of it doesn't happen here and that we're not the cause of it happening out there—possible bacteria riding along with Starman in Elon Musk's space-voyaging roadster notwithstanding.

MIXING WITH ALIEN LIFE

Cross-species or intergalactic breeding is one of the science fiction paradigms that *Rick and Morty* joyfully flips upside down—mixing human DNA with alien DNA, or possibly being infected by organisms from another world. It's played for humor in the series, but as a society, there are few things that freak us out quite as much. Think of the hazmat suits in *E.T. the Extra-Terrestrial*. What happens in *The Thing*. *Alien*. *Life*. *Invasion of the Body Snatchers*. *The Andromeda Strain*.

In real life, the Apollo astronauts had to sit in quarantine for twenty-one days after their return from the moon, specifically to see if they were infected by any microorganisms that could spread to every human being on Earth. And twenty-one days was chosen because NASA doctors, based on their experience on Earth, felt that any infection would display symptoms by twenty-one days. Seriously. We, as humans, tend to freak out at the idea of infections from space. Panspermia is a cool theory and all, and it might be true, but once life—us—is established on Earth, that's enough, thank you. We don't need any new life from space coming here, please.

If it's not a sample from an asteroid brought to Earth, the next significant infection possibility will be when we bring samples back from Mars. The idea is on the drawing board for both NASA and ESA and has been heavily studied, though we won't see any real action toward a mission until the mid to late 2020s at the earliest. Carl Sagan was one of the first to ring the warning bell about harmful microbes being in samples returned from Mars, writing in *Cosmos*: "Perhaps Martian samples can be safely returned to Earth. But I would want to be very sure before considering a returned-sample mission." More recently, many scientists have rejected the idea of bringing samples from Mars to Earth, saying that there's a greater-than-zero chance of any infectious organism escaping—and in their opinion, that is unacceptable.

Likewise, if we ever find an alien planet with natural flora and fauna, would we want to—or be able to—eat it? An alien world might have the largest, most succulent-looking fruit you've ever seen, but it could be poison to us, or utterly devoid of any nutrients that exploring humans would need. Or there could be microbes against which our bodies would have no defense. Science fiction is full of stories in which the colonizers become the alien planet's victims, but no aliens larger than a small parasite are ever seen.

Despite what movies and television show, in reality, no one would ever take their helmet off on an alien world until it has been thoroughly checked out, which would take weeks or months.

But that's not *Rick and Morty*. The series has played with the idea of interacting biochemistries in many storylines—from the gray space worms that might sterilize the planet at the start of "Vindicators 3: The Return of Worldender" to the Blim Blam who had Space AIDS in "Auto Erotic Assimilation." Not to mention the whole point of the trip in the first episode, "Pilot," was to find seeds to the "Mega Trees" to get high.

As for Rick and Morty themselves, they're constantly inviting interaction (and possibly infection) through their relationships with alien life, both casual and otherwise. Their cavalier attitude belies the notion that Rick must know far more about alien life and its compatibility with human life than he's letting on.

Throughout the series it's implied and bragged about over and over. Morty kind of has sex with an alien species in "Raising Gazorpazorp"—or at least sex with an alien robot designed to collect male genetic material in order to fertilize a Gazorpian egg. And then in "Auto Erotic Assimilation" it's virtually in our faces—Rick has sex with Unity's hosts, and it isn't anything new or even special. So, both Rick and Morty have had sex with alien species. While we're discussing sex with aliens, we should ask: Would sex with alien species produce human-alien hybrid creatures?

The answer is: maybe.

Hybrids—the term for unique offspring usually created via parents of two different species or subspecies—do occur on Earth.

It's not common, but it does happen. To create a hybrid on Earth, you've got to have overlapping territory, similar morphology, similar mating behaviors, similar times of fertility, and similar physiology. Like when a female horse and a male donkey really, really like each other, they can produce a mule. And, aside from sub-Saharan Africans, nearly all humans today carry from 1 to 4 percent Neanderthal DNA, which means that sometime in our distance past, *Homo sapiens* and *Homo neanderthalensis* mated and had offspring.

Despite many popular claims, not all hybrids are sterile. This was explained by British evolutionary biologist J. B. S. Haldane in 1922. Haldane's Rule states that in hybrid animals, the male is more likely to be reproductively viable, while the female is more likely to be sterile.

The problem of reproductive viability often comes down to chromosomes. Reproductive cells (eggs and sperm) each carry exactly half the number of chromosomes of the parent. For example, a female horse has sixty-four chromosomes (thirty-two in an egg cell), while a male donkey has sixty-two (thirty-one in a sperm cell). The mule that's the result of this union has sixty-three—an odd number that can't be split evenly for reproductive cells. Along with the numbers game, there are structural differences in the chromosomes that can prevent reproduction.

But Haldane's Rule is more of a common observation in science rather than a hard-and-fast law. As such, there are many cases of fertile hybrids, from the occasional liger (offspring of lions and tigers) to the coydog (offspring of coyotes and dogs), which is always fertile.

Going back to *Rick and Morty*, Morty Jr. was born. That means that there were enough physiological and biochemical similarities between humans and Gazorpians. In other words, the DNA in Morty's sperm was compatible with that in a Gazorpian egg (in the sex robot), and so a hybrid was produced. In reality, the chances of there being enough genetic compatibility to produce a viable hybrid human-Gazorpian would be astronomical—but this is the world of *Rick and Morty* we're talking about, so it did happen, and that's probably for the best. After all, a show where the sexual exploits of Rick and Morty with other species throughout the multiverse regularly create horrific biological consequences might not be something that would play well in Middle America. Or anywhere.

As for the question of Morty Jr.'s fertility, the Haldane Rule states that the male is more likely to be fertile. While Morty Jr. clearly had issues that he was working through in his novel, *My Horrible Father*, he is able to control the rage that's shown by other male Gazorpians. As such, Morty Jr. may just represent hope for the reintegration of male and female Gazorpians and their pathway to the future.

WHERE ARE OUR GROMFLOMITES?

Since the very first episode, the human-sized, insectoid Gromflomites have had a constant presence in *Rick and Morty*. Whether the Gromflomites built or took over the Galactic Federation may one day be revealed, but let's take a moment to look at their biology a little.

Gromflomites are clearly supposed to be man-sized insects that blend elements of common houseflies, praying mantises, ants, and perhaps even more—their name is very similar to the scientific name of the Madagascar hissing cockroach, *Gromphadorhina portentosa*. They're big and sociable and have been shown to have a sort of hierarchy—even though some may grumble about it.

Our insects are small-fry in comparison, and it's mostly due to how insects breathe. Bugs don't have lungs like mammals. In fact, you'd be hard-pressed to think of how insects handle respiration as "breathing." Instead of lungs, insects use a network of small tubes—tracheae—that allow gas exchange throughout their bodies. Oxygen comes in and diffuses across the tissues, and then carbon dioxide flows out.

Most insects don't actively move air in and out through the tracheae; it just kind of passively . . . happens, occasionally aided by movement on the insect's part. It's a good system, and the sheer diversity of insects in the world today speaks to its efficiency.

But the thing is, it's only good if you're small. If you get too big, the passive movement of air through tubes doesn't allow an effective gas exchange for all the tissues that it needs to reach. The largest insects of the modern world are the best at what they do, but they don't do it too quickly, given the limitations of their anatomy and the physics of gas diffusion. Cockroaches may get big, but truly large insects are fleetingly rare, maxing out at around 50 or 60 centimeters at the most—with a stick insect, *Phobaeticus chani*, holding the record at 62 centimeters.

But that's nothing to the record holders of the Carboniferous Period (between 360 and 299 MYA), when you could find the eight-foot-long Arthropleura millipede, dragonflies with wingspans of two feet, and more. None were standing up and making plans for a Galactic Federation, but still, they were impressive.

Insects of the Carboniferous Period didn't do anything to get around the problems with tracheae and diffusion. The world was different then. Today's atmosphere is roughly 21 percent oxygen. Thanks to the rise of ancient trees in the Carboniferous, oxygen then made up 35 percent of the atmosphere. The mechanism of respiration was the same—there was just more oxygen in the air that came in through the tracheae.

Large insects were the rule for millions of years, but the world changed. Microbes started to release more carbon dioxide into the atmosphere, which brought atmospheric oxygen down toward modern levels, and birds showed up around 150 million years ago. If you were big, juicy, and full of protein, you were dinner. Insects' average size began to shrink to the point that now, human-sized insects are only one thing in our world: nightmare fuel.

CHAPTER 2

Cloning

★ ★ ✦ ★ ★

Cloning is a given in the world of *Rick and Morty*. Even before it showed up in "Big Trouble in Little Sanchez" and again in "The ABCs of Beth," there was an understanding that, yeah, cloning was most likely something Rick had done, and, yeah, it probably took a weird turn.

But unlike a lot of the weird science showcased in the series, cloning isn't all that far removed from our world. Okay, you can't grow your own clone army or a fully grown version of your best friend from childhood who you just happened to leave behind in an alternate dimensional playground. But cloning is a well-established science. And real, actual cloning of humans may not be as far off as you think.

THE LANGUAGE OF CLONING

You may already know this, but just so we're all on the same page: a clone is an organism produced from an ancestor to which it is genetically identical. Cloning happens all the time; whole branches of the tree of life reproduce asexually—that is, instead of combining the DNA of two parents, new organisms bud or split off (with exact replicas of their predecessor's DNA) and start whole new organisms. Ever taken a cutting from a potted plant and started a new plant? Congratulations, you have a clone sitting on your windowsill.

But when we talk about cloning organisms in order to get a clone army (or perhaps a lot of copies of Rick, like in Operation Phoenix from "Big Trouble in Little Sanchez") or replacement body parts, we're talking about two different types: reproductive and therapeutic.

Reproductive cloning is what has been shown in *Rick and Morty* to date: a new animal is created that is genetically identical to the original and raised as a fully functioning organism.

Like reproductive cloning, therapeutic cloning involves the creation of a genetically identical embryo, but rather than raising it to be a functioning organism, the cloned cells remain in a dish in a lab. Cells are later removed from the clone and transplanted into the patient, to replace specific types of cells. This method has been used successfully in the lab to treat mice with a Parkinson's-like disease.

Let's dig into the science. To get both therapeutic and reproductive cloning going, the most common method is pretty simple (on paper). It's called Somatic Cell Nuclear Transfer (SCNT), and it involves surprisingly little mad-scientist-style science; just a whole lot of patient, delicate work by very rational scientists with nary a world-conquering plan among them.

To perform SCNT, you need two cells: a donor somatic cell (with a complete set of DNA—i.e., not a sperm or egg cell) from the organism you want to clone, and an egg from the female organism of the same (or a very closely related) species. The nuclear material is removed from the egg, leaving the cytoplasm behind. Meanwhile, the nucleus of the somatic cell is removed from the donor cell. The donor nucleus is inserted into the enucleated egg cell, which reprograms the host egg for the new condition it's found itself in—that is, ahem, fertilized, or what scientists call a zygote. The egg, with the DNA of the somatic cell inside it, is given a slight electrical shock to stimulate it, and it begins to divide. In a few days, the cells form a blastocyst, and later an embryo. If you're looking for reproductive cloning, you take that embryo and implant it into a female of the same (or a very close) species and let it reach term, and you've got a clone of the somatic cell-donating organism.

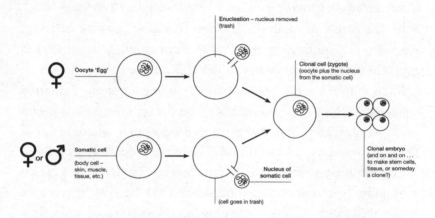

In therapeutic cloning, the goal is to get embryonic stem cells—undifferentiated cells that can differentiate into different types of cells in the body. If everything works properly, stem cells are harvested from the blastocyst or embryo and then implanted in a patient, where they develop into the types of cells that surround them. The end goal here is the stem-cell-rich embryo, not the whole organism. Once the stem cells are harvested from the embryo, the remaining cells of the embryo are destroyed or stored but not allowed to continue on their path to full development.

Terminology aside, this sounds fairly simple. But that's the thing: in theory, cloning via SCNT *sounds* easy. Just replace the instructions in the machine that can make the new living thing with new instructions, and you've got a clone, right?

Not so fast there.

"SOMETIMES, SCIENCE IS MORE ART THAN SCIENCE"

While the theory behind SCNT is sound and it seems like it would be simple to produce a clone, this method of reproductive cloning is extremely inefficient. The world sat up and took notice when the existence of Dolly the sheep—the first mammal cloned from an adult somatic cell—was announced in February 1997. After the reveal of Dolly, if the mass media were to be believed, we were just months or years away from cloning everyone and everything. The hysteria was palpable.

But Dolly wasn't a one-and-done cloning attempt. Scientists started off with 277 eggs for SCNT, and from those, 29 viable embryos were created, and of those, three were born, and only one—Dolly—survived until adulthood. Those are pretty lousy odds.

Techniques have improved, but the best current technique claims a roughly 9 percent success rate when working with species we know well, such as mice. Cloning non-domesticated animals—

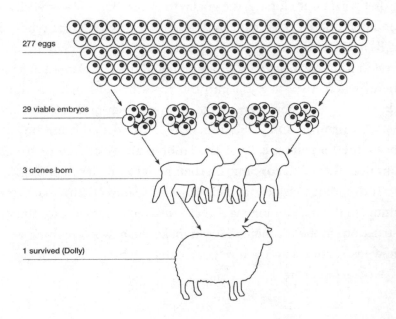

277 eggs

29 viable embryos

3 clones born

1 survived (Dolly)

that is, animals that aren't well represented in laboratories, agriculture, or homes—has a success rate of less than 1 percent. But, as technology improves, cloned organisms are often living normal life spans, and whole clonal lines of animals have survived and thrived. Researchers in Japan have produced twenty-six generations (six hundred individuals) of an original cloned mouse, and successive sheep clones after Dolly have lived normal sheep lives. But getting to the cloned animal in the first place . . . that's the trick. The difficulties along the way seem to be mountains rather than molehills.

"The major issue with these techniques is that they use somatic cells, many of which have already differentiated into specific and restricted cell fates; we're asking them to fully erase their memory and start over," explains Yelena Bernadskaya, a developmental biologist and postdoctoral researcher at New York University. "There are a number of epigenetic reprogramming barriers that scientists are still struggling to overcome. Specifically, about 20 percent of genes that are expressed in somatic cells failed to be shut off in SCNT, and about 15 percent of embryonic genes failed to turn on. Further, under regular fertilization or IVF, there are marks on the DNA carried by the sperm and the eggs that epigenetically trace the source of the DNA and distinguish them as maternally or paternally derived.

"For example, the X chromosome derived from the father will usually become deactivated so that only one copy of the X chromosome genes are expressed. In SCNT-derived embryos, both copies of the X chromosome can still be active because the signal to shut one off is missing. This results in too much gene expression that will also cause the embryo to die. It's hard to wipe the memory of a cell clean."

Some of these complications were addressed by stem-cell biologist Yi Zhang. In 2015, Zhang developed a chemical cocktail that, when added to an egg, could get stubborn genes to turn back on, leading to a tremendous increase in efficiency. Although "tremendous" in scientific terms is a bit subjective, the new process delivered blastocysts from 14 percent of adult cells that used SCNT along with Zhang's approach. Embryonic stem cells were created from 50 percent of those blastocysts.

To be clear, even though it may look as though Zhang's research has opened the door to make human cloning more likely, most researchers are interested in cloning embryos in order to harvest their stem cells—therapeutic cloning, not reproductive cloning.

That's not to say that reproductive cloning is not being performed, but it's limited mostly to agriculture and lines of experimental animals in which certain biological conditions need to be controlled from generation to generation. Reproductive cloning is also available if you've got the money—there are dozens of companies that will clone your pet from a somatic (often skin) cell, using SCNT, for the bargain price of around $50,000. Barbra Streisand caused a minor stir when she told *Variety* that two of her four dogs—Miss Violet and Miss Scarlett—are clones of her late dog Samantha. And in terms of even larger amounts of money, Texas businessman D. Alan Meeker paid top dollar to have a polo team made entirely of clones of the same horse.

With Streisand's dogs, Meeker's horses, lab mice, cattle, sheep, and a handful of others, it seems that animals are being cloned regularly. They are. The thing is, some animals are easier to clone than others.

Cats are relatively easy to clone, as are mice, for example. Dogs and rats are more difficult, but not impossible. Primates, including humans? Very, very difficult. But let's come back to that in a minute.

What makes some animals tougher to clone using SCNT than others is often the organization of the donor cell. Even with Zhang's improvements in activating genes inside the somatic DNA, if there are problems with the egg, this can cause critical issues down the road. Animals that are difficult to clone often have proteins that are needed for cell division near the nucleus of the cell. Nucleus removal from the egg is a microscopic procedure, and despite best intentions, it can get a little bull-in-a-china-shoppy. If the proteins that regulate cell division are pulled out of the egg cell along with the nucleus, the egg will never fully develop into a viable embryo, and certainly never into a Tiny Rick.

But this doesn't mean we're stuck at "now" forever. Zhang's improvements in SCNT shook the science and kick-started lines of research and cloning that were previously too difficult to pursue—namely primates. Not us, mind you, but our cousins.

"Primate SCNT is notoriously difficult for a number of reasons that we don't yet fully understand," Bernadskaya says. "After the nucleus is transferred to an enucleated oocyte, that oocyte has to be activated to form a zygote. Most commonly this is done by adding chemicals that can stimulate activation. Primate oocytes do not respond as well to the chemical cues as other species and require additional manipulation, such as applying an electric pulse, which can also damage the oocyte. Sometimes they can also activate too early, which throws off developmental timing."

Case in point: Zhong Zhong and Hua Hua, two long-tailed macaque clones that were created via SCNT in China in 2017. SCNT worked on the monkeys due to the cumulative improvements in cloning science, including better microscopes to perform the nuclear removals and better compounds for encouraging cellular reprogramming, as well as Zhang's enhancements. In the end, the method used to clone the monkeys was effectively the same as the method that created Dolly—with twenty-one years' worth of new science along for the ride. But with all the new science and technology, the Chinese method was far from efficient—it reportedly took 63 surrogate mothers and 417 eggs to produce four clones, two of which died shortly after birth, leaving Zhong Zhong and Hua Hua as the two success stories.

But we can expect this science to continue to improve. "One

way that scientists are considering increasing efficiency is through the use of CRISPR," Bernadskaya says. "CRISPR is a technique that can do a number of different things, such as cut DNA strands at directed locations, recruit protein factors that can block expression of a target gene, or force expression of some genes. We may be able to use CRISPR to shut down some genes that have been hard to shut off automatically in SCNT and turn on some others, hopefully increasing the ability of the embryos to survive and implant."

JUST BECAUSE WE CAN DO IT, SHOULD WE DO IT?

Without getting into the murk of cloning ethics (and there's a lot of murk), the inefficiency of producing new clones is one of the many practical issues that keeps human cloning off the table. If a company cloning cattle or pets has an insanely high 10 percent success rate, that's one thing. Sure, there were lots of embryos, fetal animals that never made it to a live birth, and even animals that were born but died shortly after—but they weren't . . . us. A 10 percent success rate for human cloning would mean a very large number of failed human embryos and/or pregnancies.

Yi Zhang emphasized this point when news broke of his advances, saying that no society would accept the low efficiency and loss of embryonic human life.

And Zhang's not wrong. On the record, no one is thinking of using any of the methods laid out here to clone humans. Cloning—of any kind—is banned outright in forty-six countries, while reproductive cloning is specifically banned in thirty-two others. Many countries leave the cloning door open by allowing therapeutic cloning, but given that the technique is also associated with the creation of embryos, therapeutic cloning for human treatment is still a sticky subject in many places. There are hold-outs, of course—China has not formally banned cloning—but let's be clear: that doesn't mean that they are cloning humans.

Putting ethics, technology, and science aside for a moment, we can consider how cloning humans would take a tremendous amount of money and resources. Millions, perhaps billions of dollars will be required to get the science and technology to where it needs to be—not to mention thousands of human donor eggs.

Objectively: the efficiency of cloning has improved greatly since Dolly. It will continue to improve. Eventually, from a viewpoint of science and technology, human cloning will be possible. It's just not there yet.

WHERE DO YOU KEEP YOUR CLONES?

Let's say you've figured out a way to make cloning work and, like Rick, you want to set up your own Operation Phoenix in your basement as a fallback. Or maybe you want to grow a clone of a childhood friend and need somewhere to do it.

Unless you've found a very cooperative woman, you're going to need some artificial wombs in which the clones can develop. This is called ectogenesis, or development ex utero—gestation outside of a biological womb—and research on it has been going on for decades. In science fiction, including *Rick and Morty*, these artificial wombs are often shown as large jars, but in reality, they're more like big plastic bags.

In 2017, a team led by fetal surgeon Alan Flake unveiled the Biobag—an external womb that's designed with the hope of giving premature infants a more natural environment in which they can continue their development. In initial tests with fetal lambs that were between 100 and 120 days into natural gestation (the equivalent of about twenty-seven weeks in human pregnancy) before being placed in their Biobags, the artificial wombs performed almost as well as the real thing, keeping the animals alive for up to four weeks. For the Biobags, the team developed an artificial circulatory system without a pump, as well as a unique oxygenator, which allowed each fetal lamb's developing heart to pump blood through its circulatory system. The system also featured circulating synthetic amniotic fluid. Altogether, the Biobag system looks like a baby lamb inside a large Ziploc bag. While in the womb, the animal is connected to the outside world by tubes through its umbilical cord, and that's it. After time, all the lambs in the experiment wriggled, opened their eyes, and even started growing wool. The animals were removed from their bags after four weeks, and while some were euthanized for further study, some were allowed to develop, and appeared by all measures to be normal lambs.

"This particular technology is not aimed at growing animals to full term," Bernadskaya says, "but rather as a support system for premature babies to provide a more natural environment for their development. It's been suggested that human trials of the Biobag can start within the next three to five years. But while we might be close to having a therapeutic tool, the ability to grow a human fetus to term is still pretty far off."

Operation Phoenix isn't full of babies—which would have been horrible, given how Rick destroyed his cloning plan—but rather of mostly full-grown Ricks. In our world, that would mean that the clones have been in their chambers for a long time. Years. Raising a kid outside a clone bank is pretty pricey, and while clone kids won't want to play soccer, growing a clone of yourself from embryo to adult would require a significant outlay of resources, including a whole lot of money. You'd need specialized fluids, filters, ways to maintain the clone, and more. The cost of growing a clone to adulthood—using our science as a metric—would be prohibitive.

CLONES AREN'T ALL THAT

A batch of pre-grown clones ready at a moment's notice sounds great on paper, and things seem to be working out well in the areas of agriculture, polo teams, and celebrity pet resurrections, but again . . . there are pitfalls.

Firstly, your clone is you, warts and all (well, technically, your clone won't have the same warts as you). All genetically based problems are going to be repeated in your clone. All of them. There's no genetic variability. If, for instance, your DNA includes a gene that always results in the development of a disease later in life, your clones carry that same ticking time bomb. It's the same for any vulnerability you may have to other diseases, whether they come from viruses, bacteria, or another vector. This can be a major problem, for instance, if you've invested millions in a herd of identical beef cattle and you watch them all succumb to the same disease, because when it comes down to it, they're all the same animal. Three cheers for sexual reproduction here—two batches of genetic material becoming one helps out greatly with genetic diversity and hybrid vigor, which add beautiful differences to a population.

Secondly, your clone is not you. Hold on—how about we put it this way: your clone is not "you." Pull a fresh version of you out of your clone tank, and they're not going to pick up your sentences where you leave off like some kind of telepathic twin. Unless you've copied your consciousness into the clone via some unknown technology, they're a blank slate. They've had no experiences and know, quite literally, nothing. Even if they do manage to know something, their experiences are different from yours, and they'll be, at best, a twin brother or sister rather than a copy of you. It's nature vs. nurture—your clone will have a completely different "nurture" experience, despite having an identical "nature."

And even if you could swap your consciousness into the brain

of a fresh clone, you're going to have problems. It's what "Big Trouble in Little Sanchez" played with so brilliantly: putting an older man's consciousness into a brain that is physically much younger is going to be troublesome, to say the least. The architecture of Big Rick's brain is significantly different from that of Tiny Rick's. Jokes about how an older man would function in today's teenage culture aside, for someone as aware of his own intelligence and brain function as Rick seems to be, having all of that crammed into a brain that wasn't working as well would be a form of torture (probably something bad enough to make you write songs and do dances about wanting to escape from your body and how you're dying in a vat in the garage).

"During adolescent development our brains undergo extensive pruning and rearrangement, potentially linked to our experiences and interactions," Bernadskaya says. "Too little or too much pruning can have serious consequences later in life, and have been linked to some forms of schizophrenia. Since the Ricks in his clone bank are developing way past adolescence and are having no experiences, chances are that they would come out of the tubes with severe mental or behavioral disabilities."

CLONE MYTHBUSTING: AGING A CLONE

In "The ABCs of Beth," Rick and Beth had a wonderful father-daughter moment when they cloned and regrew Beth's best childhood friend, Tommy, from the DNA extracted from his finger. It took about three hours and involved a microwave with Rick-style adjustments and the song "Fathers and Daughters," best known for its catchy line about having a "doo-doo in my butt."

Three hours from DNA to fully grown is extreme, but explains how Rick has a bank of clones, as well as his offer to Beth at the end of the episode to make a clone of her that would replace her in the Smith household—which she may or may not have taken him up on.

While modern science can regrow a limited number of organs to functional adult size in a relatively short amount of time, there's nothing that can speed up the clock for an entire human, and to even think of doing so is fraught with difficulties. As mentioned earlier, mental development may very well depend upon experiences and interactions over time to produce a healthy brain. Hitting fast-forward on human development would likely cause many more problems than it would address.

A tragic form of this speeding up of the clock is seen in the disease progeria, whose victims eerily appear to be fifty to seventy years older than their actual age. Aging is not progressive, and the victims of the disease move rapidly from young to elderly and often die from diseases normally seen in individuals much, much older than they are. The causes of progeria are genetic, and there is no cure.

So, aging a clone in a vat is still limited to Rick's science, not ours.

But the topic of aging a clone does give us the opportunity to shoot down a clone myth—that Dolly and other clones are somehow "born old" or born as "old" as the cell from which they were cloned.

This idea got its legs when Dolly the sheep lived for only six years after being born. Her breed has a normal life span of twelve years. At the time of her death in 2003, it was reported that Dolly was "born old," since the somatic cell that she grew from was six years old at the time of successful cloning. Six years alive after being born from a six-year-old cell equals twelve years. The math was easy, and it looked like a smoking gun had been found.

Oh, and by the way, the somatic cell that was used to create Dolly came from the mammary gland of another sheep, and her name was chosen in honor of country singer Dolly Parton. This has nothing to do with the science of cloning, but it is something that may get you a chuckle at a party.

When a cell divides through the process of mitosis, genetic material is bundled into organizational structures called chromosomes. At the end of each chromosome is a telomere—think of it as a "cap." During replication of the chromosome, a complete copy isn't made; with each progressive generation of chromosomes, some of that telomeric "cap" is shaved off.

So telomeres seem to decrease in length over time. An older individual will have shorter telomeres than a younger one. "Older" cells act differently than "younger" ones do, as well. Again, to bring this back to reproductive cloning, Dolly was cloned from mammary glands that were six years old at the time of cloning, so it's not too surprising that she lived for only six more years, or so the claims went.

But that's not what's being seen when the data is looked at carefully. Along with other sheep from her herd, Dolly died from a fairly common form of lung cancer caused by a virus, which was exacerbated by her being kept indoors (Dolly was kept inside for security reasons). Further study of Dolly's tissues showed no evidence of premature aging. Dolly was, effectively, six years old.

"While it is possible for the telomeres to be shorter in cloned animals compared to those in normally conceived or IVF animals, there seems to be a lot of variability in telomere length between individuals to begin with, making it hard to associate telomere length with physiological problems of the clones," Bernadskaya says. "In the oocytes, an enzyme called telomerase ensures that telomeres do not shorten prematurely. This enzyme can work on the telomeres of the donor nucleus, thereby preventing more damage. Generally, with improved SCNT techniques, the life span of cloned animals is approaching that of normally conceived ones."

CHAPTER 3

Ecology

★ ★ ✦ ★ ★

In the three seasons to date of *Rick and Morty*, there have been shout-outs to ecological topics such as invasive species, energy flow within larger ecosystems, and the possibility of ecosystem collapse, and a ton of bizarre life-forms that just "fit" in their respective worlds, tucked inside their particular ecosystems.

For a show about a crazy scientist and his anxious grandson, *Rick and Morty* gives more than a few nods to ecology.

Ecology is a broad area of science that looks at how living things interact both with one another and with their environments. First, some vocabulary: living organisms and the things they produce, or the resultant actions thereof, are called biotic factors; nonliving things and their actions are called abiotic factors. A lion eating a gazelle or a bacterium working to decompose a log and release its nutrients—both biotic. An asteroid crashing into the planet and ruining your big, dinosaur day—abiotic.

As a science, ecology comes with a terrific organizational hierarchy, from smallest to largest:

* **Organism:** the individual living thing in an environment.
* **Population:** a group of individuals, all members of the same species.
* **Community:** all the populations in an area that interact with one another (that is, all the living organisms, or all the biotic factors).

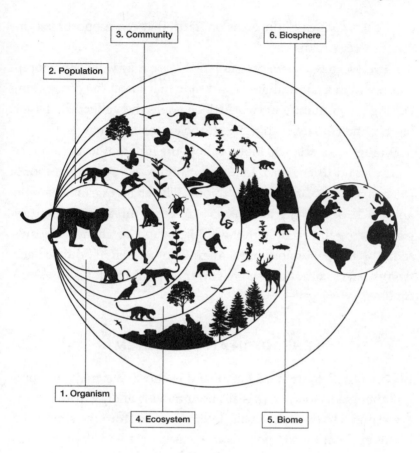

* **Ecosystem:** a community plus the associated abiotic factors of the area in which it exists.
* **Biome:** a collection of similar ecosystems.
* **Biosphere:** the portion of the planet on which life is found.

The first three—organism, population, and community—can all exist inside the same forest, for example. The ecosystem is the community along with the abiotic factors of the forest, such as the soil, terrain, air quality, and water availability. All the similar forest ecosystems make up a biome, and the biosphere, in this case, encompasses all the areas on Earth that can support life (forests, deserts, oceans, etc.).

Each level has its own pressures: a disease may wipe out a population within a community; a rockslide may affect the ecosystem, cutting a community in two—which could then affect the larger biome; and climate change affects the entire biosphere.

Ecology, as a science, looks at how the different levels interact and respond to changes. Like living things, ecosystems respond to changes, whether biotic or abiotic. The changes can be simple and quick, or complex and massive in scope, threatening the very existence of the ecosystem itself. Life adapts to change. Ecosystems, biomes, and biospheres are resilient and often bounce back from stresses, although we are currently putting a lot of pressure on ours.

ENERGY IN AN ECOSYSTEM

Life in any ecosystem on Earth, and in those that may be found on other planets someday, is all about moving energy from the energy source to the individuals. Energy comes from the sun or the bonds in chemical compounds and is used or stored by organisms. On Earth, the most common example of this is sunlight hitting plants and the plants capturing a small portion of that energy to produce sugars, which store it (although there are creatures that can get energy directly from chemicals themselves). Organisms that can take the energy from the source and store it are called producers.

After the energy is stored, some of it can be transferred to the organisms that eat the producers—the primary consumers. On Earth, these are most commonly herbivores, such as insects, livestock, and vegans: they receive their nutrients and energy directly from plant-based material.

Existence gets dangerous from here: secondary consumers and tertiary consumers show up to get their energy and nutrients from the level below—either by consuming the consumers or their by-products. These are called trophic levels, and taken together, they form a pyramid, with producers forming a wide base at the bottom and the top consumer, or apex predator, at the top. The older or more established an ecosystem is, the more efficient the energy flow upward, thanks in large part to the biodiversity throughout the entire system.

This pyramidal shape is pretty much how we would expect to find life throughout the universe. Each successive level gets smaller and smaller as it feeds on the level below it. While you can have an ecosystem with a massive number of producers and primary consumers and few predators, you can't have a planet full of apex predators without producers and a layer or two of consumers. Likewise, you can't have apex-apex predators or an indefinitely high pyramid. There's just not enough energy to go up.

The energy flow between the levels, frankly, sucks in terms of productivity, but that's what happens with the first two laws of thermodynamics. The first law states that energy cannot be created or destroyed, only transformed from one type into another (this is also known as the law of conservation of energy). The second law says that when energy is transformed from one form into another, some energy will always be lost, almost always as heat.

Taking this back to the trophic pyramid, this means that only about 10 percent of the original energy taken in by the producers (plants, for most of us) makes it to the next level. The energy that doesn't make it to the next level is used by the plant to build and maintain its tissues and structures, as well as to fuel its metabolism, respiration, and all the other needs associated with being alive. This pattern is repeated at each level of the pyramid: for example, mammals and birds use a lot of energy for locomotion and maintaining body temperature and, in addition, a large portion of the energy taken in by an animal is "locked up" within the organism, as with plants. Teeth and bones, for example, represent energy that cannot move upward in a trophic pyramid, and their energy will be available only to decomposers after the organism dies, if at all.

Primary consumers (cows, for example) would get 10 percent of the original energy that plants gathered from the sun, and then humans (from eating the cows) would get a measly 1 percent of the plant's energy. If you're eaten by a lion after chomping your burger, the lion would only get 0.1 percent of the original energy that was captured by the plants that were eaten by the cow, which was eaten by you. Add a lion-eating super-predator (an apex predator, above lions, with no natural predators of its own), and that would get even less of the energy. Most food chains illustrated by trophic pyramids can support only four to six levels, with the ocean's pyramids edging out those on land for number of layers. There's just not enough energy for more.

As a side note: don't hate on thermodynamics. The laws are just what they are. Instead, focus on the plants, as they are at the RIPE Project at the University of Illinois at Urbana-Champaign. This project, like many others being worked on around the world, is looking at tweaking the efficiency of the photosynthesis performed by plants on that bottom layer of the pyramid. Of the energy that reaches them from the sun, plants turn only about 5 to 10 percent of that energy into biomass that can be utilized higher up the chain. Through genetic engineering, the hope is to increase that percentage so that more energy would be converted into biomass, therefore making the plants better food sources for the primary consumers.

Still, the changes being studied by the RIPE Project and others won't completely upend trophic pyramids and ecosystem energy flows—you can't fight thermodynamics' second law—but it could bump the energy up a little, which is something we're going to need to do in the face of a changing climate and reduced biodiversity.

BALANCED ON THE HEAD OF A PIN

Ecosystems grow and change, depending upon the conditions in which they exist. Established ecosystems that have existed for centuries or longer can be highly efficient and complex (think of the Amazon rainforest), while others are so delicate that one small perturbation could upend the entire thing. But whether the ecosystem can roll with the punches or is ready to collapse if one individual organism catches a cold, they all are, in their own ways, fragile.

Ecological or ecosystem collapse can be triggered by either abiotic factors (asteroids, volcanoes, or climate change—either abrupt or over time) or biotic factors (overfishing or predation, habitat destruction, or invasive species introduction). If the disturbance is large enough, the entire ecosystem will suffer a reduction in "carrying capacity"—that is, the number of individuals that can survive in it. In the extreme, this can lead to mass extinctions in which entire populations or species die out. Don't lose focus on the bigger picture, though: to quote Jurassic Park's Dr. Ian Malcolm, "Life, uh, finds a way." Ecosystems have and will continuously supplant one another on Earth. Whether or not the resultant ecosystems can support human life, and the bulk of life on Earth as we know it—that's the question.

And the creators of *Rick and Morty* seem to know this, even if they show it in many unusual ways.

RICK AND MORTY'S ECOLOGY

Threats to ecosystems have played both major and minor roles in *Rick and Morty*, throughout the course of the show's first three seasons. After all, on the resort planet of "The Whirly Dirly Conspiracy," Rick explains to Jerry that he needed him coated in gibble snake bile in order to attract a shmooglite runner. That's a quick exploration of some predator-prey relationships between trophic levels right there.

But on the serious side: on many occasions, the series's writers have shown that even a minor change to an ecosystem can be enough to turn things horrible, and even make evacuation of the entire reality necessary.

HIT THAT ECOSYSTEM . . . HARD

In terms of major changes, Rick upset the entire C-137 biosphere in "Rick Potion #9," or what we could call "Dan Harmon & Justin Roiland's Modern Frankenstein." Think of the warning that was baked into Mary Shelley's *Frankenstein*: when she wrote it, humans were reaching farther than we ever had, and the idea that we, as a species, were trying to become—or play—god, or spit in god's eye, was an issue of serious philosophical debate. This was all at a time when sciences such as chemistry, biology, and medicine were starting to intrude on regions once thought to be the exclusive territory of the divine. Thanks in part to Shelley, the question "We can, but should we?" is one that has troubled us since the earliest days of modern science and medicine. What happens when we overreach, and what are the unintended consequences that can happen when we do?

"Rick Potion #9" starts with a simple premise: Can Rick whip up some science to make Jessica fall for Morty? Rick's initial potion, which Morty smears on Jessica, contains oxytocin from a vole and

Morty's DNA, and uses some old-school genetic modification by mixing ingredients (no gene-by-gene editing via CRISPR needed). But then things go sideways when the potion piggybacks on the flu virus from Jessica. Rick's plan to fix it? Just throw some praying mantis genes and a more contagious flu virus together for an antidote. Unfortunately, that causes the massive morphological changes that Rick so politely calls "Cronenberging," after the pretty horrifying special effects of director David Cronenberg's horror films.

What this did is . . .

Screw.

The.

Ecosystem.

Specifically, the big problem was that the world's population changed their entire reproductive drive. Reproduction (along with turning people into mantis-human hybrids) was now focused on one individual rather than being spread throughout the respective populations, and as a result, ecosystems, biomes, and, ultimately, the whole biosphere collapsed.

You see the full effects of this in "The Rickshank Rickdemption" when Summer and Morty return to Earth C-137 to see that almost all life has vanished, and the remains of human society—apparently, just the Smiths—live in a postapocalyptic wasteland: no vegetation, the sky is not quite the right color, and the only thing to eat is "roasted Cronenberg." Things apparently got bad—really bad—after Rick and Morty left. It would have been one ecosystem collapse after another after another as the planet worked to find a new equilibrium. And if "Hunger Games Summer," Beth, and Jerry are the only humans left in C-137, the new equilibrium that the ecosystems of Earth in C-137 reach most likely won't include them. There just aren't enough individuals to allow for a healthy genetic diversity and a population that grows over time.

Like *Frankenstein*, "Rick Potion #9" can be seen as a (pretty extreme) parable about messing with things we don't understand. Rick himself admitted that he made a mistake in the engineering of the potion, the antidote, and his final concoction (that included dinosaur and Golden Lab DNA) that was supposed to revert things back to normal, but was probably going to screw things up even more.

INVASIVE SPECIES . . . FROM SPACE!

Rick and Morty has dealt with species that don't belong in an eco-sysystem in at least three episodes, and each time, things went wrong (or were headed there in a hurry). An invasive species is an organism that's not native to the region it's in, and, as a result, can easily upset the ecosystem in any number of ways. Technically, invasive species are ones that adapt to a new area and reproduce quickly, with the potential to out-compete native species for resources within a community, ecosystem, or biome. Invasive species are also called "alien" species, which for our purposes works perfectly.

Any extraterrestrial life that shows up on Earth in *Rick and Morty* is an alien species and has the potential to change the biome, community, and ecosystem into which it is introduced. From the synthetic laser eels of "Ricksy Business" to the one-eyed blue worms at the start of "Vindicators 3: The Return of Worldender" with the potential to sterilize the human species to the bug on the windshield at the beginning of "Look Who's Purging Now" to the telepathic alien parasites of "Total Rickall" and even, to an extent, Unity; all these invasive species changed (or had the potential to change) the ecosystem in which they found themselves.

A lot of invasive species on Earth make news when they damage property (zebra mussels cost communities near the Great Lakes tens to hundreds of thousands of dollars annually in cleanup and replacement of mussel-encrusted facilities) or the economy (agriculture around the globe loses billions of dollars annually to invasive species control and prevention, and their very presence has an upsetting effect on the ecosystem they move into).

Think back to the trophic levels in the pyramid: destabilize one of those levels and the whole structure is threatened. Increase predation on a primary consumer or increase consumption of the primary producers so the native primary consumers can't get as much to eat, and the whole thing starts to teeter. If it teeters enough, the ecosystem could collapse. And while the collapse would eventually result in the formation of a new (albeit less diverse) ecosystem, that could upset the larger biome.

But it doesn't always teeter over. While "invasive species = bad" used to be the accepted orthodoxy on the subject, many ecologists around the world are now calling for a more nuanced view of nonnative species, especially considering the extent of habitat destruction and other environmental changes due to humans, including climate change. While negative effects of invasive species are easy to find, there can be benefits: monarch butterflies nest in nonnative eucalyptus trees in California; the Spanish reintroduced horses to the Americas after they had gone extinct there; honeybees, honored by several states in the United States as their "state insect," are, in fact, immigrants to North America, originally starting out in eastern tropical Africa; and, in several instances, people are looking at ways to eat themselves out of invasive species issues by turning the nonnative species into desirable food sources.

Still, "concern" is probably the correct term when it comes to invasive species, given that any nonnative species will affect the native ecosystem. Whether that impact will be large or small depends upon many factors.

But let's all agree to stop the one-eyed blue worms before they sterilize the planet.

ECOLOGY OF THE VERY SMALL

It's worth bearing in mind that ecosystems don't have to be expansive; they can also be very small. Your body, in fact, is home to many different microbiomes. These microbiomes are organizationally similar to what you find in the macro world—there are individual bacteria and other microbes that make up communities and ecosystems, each with their own different biotic and abiotic factors. These different ecosystems are home to a wide variety of organisms, some that thrive in the presence of oxygen and light (which can be found on the skin), and others that crave less light, more moisture, and lower oxygen content (such as inside the mouth).

Or inside the body—as was shown in "Anatomy Park." Like the biomes and communities of the Earth, Reuben's health went downhill fast when his internal microbiomes were upset by sabotage. When the ecosystem in Reuben's lungs was shifted just a little, the tuberculosis that had been dormant came roaring back to life, causing the acute respiratory distress that ultimately killed him.

While *Rick and Morty* isn't at the forefront of medical research, the larger idea of the involvement of microbiomes in the human body in overall health is an active avenue of research. There are many microbiomes in the body—no part of the human body is sterile—and a lot of current research focuses on the gut microbiome, where a diverse collection of bacteria and other microbe species live.

While the microbiome of the gut didn't play a huge role in "Anatomy Park," that's not to say that some preventative work on Reuben's gut microbiome couldn't have positively affected his overall health. Research suggests that individuals with chronic diseases such as type 2 diabetes may have a different makeup of gut biome species than those without. While studies on the benefits of taking probiotics (microorganisms taken orally, with the intention of positively affecting the intestinal biome) for overall health are inconclusive, fecal microbiota transplantation (via orally administered "poop pills" or direct fecal transplant) is showing success in treating patients with chronic or recurrent intestinal infections of harmful bacteria. Maybe with the new Anatomy Park going up in Ethan, the happy Small World–style ride should be located in the intestines.

Our microbiomes are no minor thing, either. Nonhuman cells in our bodies outnumber human cells on the order of 10 to 1. Most people have between two and six pounds of nonhuman life inside them at any given time, which can certainly add more scientific weight to the ageless philosophical question "Who am I?" If "you" make up only around 10 percent of the cells of "you," then who's calling the shots? Before answering that, consider that there is evidence that the gut microbiome can influence behavior and perhaps even thought.

Regardless of who and what we are, microbiomes are present in virtually every corner of your body—yes, even down there . . . and definitely up there. Upset their balance and, like other communities, biomes, and ecosystems, they will struggle to regain some form of equilibrium. And in the case of microbiomes, that new equilibrium may have short- or long-term effects on your health.

And that's the larger picture here—along with the message coming through from *Rick and Morty* that big things, small things, and their associated unintended consequences can change an ecosystem in very bad ways. Of course, in the grand scope of things, the Earth will, ultimately, be okay. If an ecosystem goes down, a new one will develop to take its place.

But for the record: we should probably be working to make sure that any future ecosystems on the planet include human beings.

KNOW YOUR E. COLI

E. coli, or *Escherichia coli*, is a type of bacteria that we most often hear about with food contamination, with said contamination most often resulting from the exposure of the food to fecal material. To talk bacterial science for a moment: there are many strains of E. coli bacteria, and all of them are classified as being facultative anaerobic. This means that they can live both with or without oxygen to perform the necessary functions of life, such as respiration, the process by which they create ATP, which is used as energy.

The bacteria is commonly found in the lower intestine of warm-blooded mammals, right where Morty and Anatomy Park's survivors found it on their way out of Reuben via the lower intestine. Normally, the E. coli that inhabit the gut are harmless, and help the host by producing vitamin K2, which is required for protein synthesis and moderating the populations of harmful bacteria. Additionally, E. coli is widely used in research, thanks to the ease with which its genetic material can be modified and its rapid reproduction rate—twenty minutes if conditions are favorable.

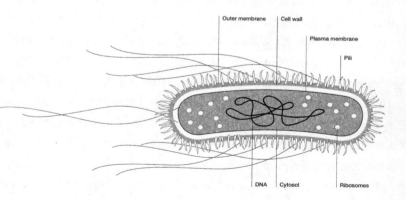

Once outside the body—it exits via fecal matter—E. coli goes fully aerobic and reproduces copiously, thanks to the oxygen in the air. This is when contamination usually occurs. Aerated fecal matter with harmful E. coli makes its way into a segment of the food supply pipeline, where, if it isn't caught, it can infect consumers and cause light food poisoning at the mildest to Crohn's disease at the most severe.

The appearance of the E. coli outbreak at the Bone Train station inside Reuben was probably meant as a visual metaphor that this was a dangerous strain of E. coli, but in reality, E. coli belong to a classification of bacteria called bacillus or bacilliform, meaning that they are rod-shaped. An E. coli colony looks like a pile of elongated barrels, all clinging together. And they're not that mobile.

The E. coli in the E. coli outbreak look an awful lot like viruses called bacteriophages. These things look like straight-up monsters from science fiction. To operate, the bacteriophage "lands" on a bacterium with its fibrous "legs," pushes a tube through the bacteria's outer membrane, and injects its own DNA into the bacteria. The bacteria takes up the phage's DNA, incorporates it, and then, zombielike, starts creating new phage pieces, which are then assembled to create new bacteriophages. The host cell eventually splits open or is cut open by the phages inside and spills out tons of new phages, which then go on to infect other bacterial cells.

Despite their horror-show appearance and operation, bacteriophages aren't all terrible. Effectively, they're small biomachines that can infect bacteria with genetic material. As such, bacteriophages are a hot topic for research as a means to alter bacterial DNA and to attack and kill specific bacteria, such as bacterial agents used as bioweapons, bacterial infections of food materials, and therapy for drug-resistant bacterial infections.

RICK'S MICROVERSE—TERRIBLE IDEA, OR LITERALLY THE WORST IDEA EVER?

Let's think about an ecosystem that's bigger, or smaller but working for something bigger. A whole world that powers another . . . thing. A world that, in a way, is just part of a trophic level in a larger pyramid.

Rick's microverse battery in "The Ricks Must Be Crazy" should be providing 20 terawatts (1 terawatt = 1 trillion watts) of juice to the engine of his car, but instead, it's sending zero at the start of the episode. Rick and Morty go to investigate. Got it.

Rick made the microverse by putting a spatially tessellated void inside a modified temporal field, until a planet developed intelligent life. He got the planet hooked on the idea of electricity, which they produce at an astounding rate (more on that in a minute), sending 20 TW up to his car's engine to run it. While a direct conversion isn't perhaps the best way to use electrical measurements, an average car battery has about 4,000 to 5,000 watts.

Rick's microverse world is more advanced than Earth, and that comes with some safe assumptions we can make about energy. Their energy production, which comes from stomping on the Gooble Box, must be efficient, so the energy lost in the transformation (mechanical energy of movement into electrical energy) must be pretty minimal. It's not a suspension of the Second Law of Thermodynamics, but probably a Rick-informed push of it to one side for the duration.

Twenty TW is the energy output from the planet . . . but how much must be produced overall if the world can constantly have 20 TW ready as soon as Rick's key turns in the ignition, without a loss of power on the planet? Again, the Gooble Box is probably efficient, but the planet also needs its own energy. It's a hugely technological society, so their energy needs would be at least as much as Earth's, which is around 20 TW-years, or 175,320 TW-hours (1 TW-year = 8,766 TW-hours), and each of those TW-hours is 1 trillion watts consumed per hour, planetwide.

I say "at least" because of their technology and their advancement. Given their technological advancement, they have probably figured out how to be more efficient with their energy production, use, and transmission than we are; sorry to say, humans are still pretty lousy on this point. The average efficiency of a natural-gas power plant is about 40 percent. That means that for all the energy released in the plant's operations, only 40 percent comes out as useful electricity. Plus, there's more loss in the transmission to your home and in your electronic devices as well.

In short, the population would have to have 20 TW in the energy "bank," ready to go at a moment's notice, anytime Rick started the car, in addition to all the energy needed to run the planet. All generated by people stepping on their Gooble Boxes. A 100 percent efficiency exchange doesn't exist—thermodynamics says so—so the people of the microverse would have to create more energy than that required to fulfill Rick's needs. So much more.

The total amount of energy produced by the planet would easily be triple or quadruple Rick's needs. They need to produce that much, though, just to always be ready while somehow keeping their planet running.

No wonder Zeep Xanflorp created his own miniverse to power their world. That was an insane amount of energy.

But here's the thing: let's look at this in Earth and solar-system terms. The sun pumps out 4.10×10^{14} TW in all directions, and the Earth gets 174,000 TW of that. If the planet in the microverse has a similar star, is there no way to capture some or all of that energy?

In the microverse, Rick created stars and waited for a planet around one of those stars to evolve intelligent life, and then he told them how to create electricity via Gooble Boxes. The people had the technology to make electricity via the Gooble Boxes, but apparently they didn't have the science needed to see a better way to make electricity, or at least not until this episode. So they stuck with the Gooble Boxes.

But back up—think about their star, and the millions or billions of other stars like it in the microverse. They were putting out more than enough energy to start Rick's car, and to run the A/C at full blast while cranking out music on the sound system. Rick knew that, but instead he chose to create a whole planet, essentially enslave the population to create a massive amount of energy for him (he got only a small percentage of the total amount they needed to produce—think trophic levels and thermodynamics), and not give them the tools to find an improved way to do it. Think of it as though plants were sentient, and even as we benefit from their energy production, we know a better way for them to get more energy but don't bother to tell them about it.

It's no wonder Morty spends the entire episode pissed at Rick. It's a pretty horrible way to get energy for his car.

CHAPTER 4

Evolution

★ ✦ ✦ ★ ★

Before we get started: you're reading a book about the science of an animated series. Spoiler: there's been a bunch of science so far, and there's going to be more. That said, the next one should come as no surprise: evolution is real. Evolution has been observed in the lab, in nature, and across the millions of years of Earth's history. It has a tremendous amount of facts to back it up. If you feel otherwise, 1) you're really going to have problems with the chapter that deals with the creation of the universe, and 2) you may want to put this book down now.

As a science educator, I just had to say that. Moving along.

When Charles Darwin was considering the mechanics of evolution, two things he probably never thought of were how alien life would evolve on alien planets to produce creatures adapted to conditions unlike anything on Earth, and whether evolution would proceed the same way if it happened under the exact same conditions in multiple realities. Darwin was all about finches and tortoises and couldn't give much time to worrying about alien life and alternate realities.

I feel pretty safe in my stance about those.

Evolution can be most simply described as change in organisms over time. The change comes as a result of random variation in traits that are passed from parent to offspring—that is, that are inheritable. If the traits help the organism both survive predation and out-compete other organisms for resources, then those organisms get to have babies and pass the genes responsible for the traits along to the next generation. Over time, those beneficial traits and their associated genes increase in frequency among a population.

Think of our distant ancestors: if you're reading this now, your ancestors were most likely pretty good at hunting and could recognize patterns, like large predators coming through the grasses. Early humans who, for whatever reason, could not recognize the shifting patterns in the grass that indicated a predator was coming are no one's ancestors. They didn't get to pass along their genes.

This, in a nutshell, is natural selection: populations of organisms become better and better suited to their environments over time. But no environment lasts forever. If the environment changes, natural selection will favor organisms with genetic mutations that are different from the ones that preceded them, with the new organisms perhaps better suited to the environment than the old.

Given our understanding of how life works, evolution is universal—that's not really up for debate among evolutionary biologists. But there are two large schools of thought on what life that evolved on an alien world would look like: Earthling-like and human-ish (i.e., evolution is convergent), or not really Earthling-like or human-ish at all, which promotes the idea that life will always show up, commonly referred to as "contingency." Each side has their reasons and champions. Convergence is often seen as a view of evolution that opens the door to a "designer" or intelligence that is, or has, guided the process, leading it to its ultimate (in our case, humanoid) end point. Contingency is the view that evolution occurs, by and large, through a series of random occurrences and events, no designer or intelligence needed.

Like lots of science fiction, *Rick and Morty* tends to lean heavily on the idea of humanoid aliens, with only an occasional nonhuman body form (think Fart from "Mortynight Run"). While you can argue that the prevalence of bipedal, upright aliens is a projection of our own humanoid biases, there is a line of science-based reasoning that suggests a universe of humanoid aliens is more than wishful thinking.

Championed by evolutionary biologist Simon Conway Morris and others, convergence in evolution suggests that, given similar challenges from environments, life would ultimately seek out and utilize similar solutions. Many examples of convergent evolution exist on Earth, where animals from different evolutionary branches have developed similar adaptations for the same environments. Dolphins and sharks have similar body shapes and means of propulsion through the water of their environment, but are not closely related. Marsupial opossums and New World monkeys both have prehensile tails. Human and octopus eyes are uncannily similar. African and North American porcupines, despite their obvious similarities, are two separate species, with no common evolutionary connection, and appear to have evolved independently.

And there are plenty of examples of convergent evolution in unrelated species when you look at analogous structures. For example, flying animals, from birds to bats to butterflies, all fly using the same basic mechanics and structures. Conway Morris even points out that convergence can be seen in societies—both ants and humans have discovered agriculture, and if you look at the tree of life, we're very far apart.

All things considered, proponents of convergent evolution reason that the examples of life as we see it on Earth point to the fact that all life forms will share similar forms, parts, and functions, up to and including the "inevitability" of a humanoid body type throughout the universe. The thought is that as far as we understand, the planets throughout the universe share the same physical laws, and life has the same overall goal. Flowering plants are fairly sophisticated answers to the problem of getting energy from your local star and making sure your genes are passed on to future generations; likewise flying, swimming, and being warm-blooded, among other challenges in animals. There are only a handful of pathways down which life can travel—and many of them wind up at a humanoid destination. As a result, life's solutions to environmental challenges resemble one another, and we can extrapolate that life on Earth-like planets should be similar to Earthlings, which is mostly what's seen in *Rick and Morty*.

However, there is another side to the argument, as evolutionary biologist and author of *Improbable Destinies* Jonathan Losos told me. Contingency is the theory that the evolution we see on Earth is just one of many unique pathways to reach the same goal, and the reason that we see the diversity of life that we do is because evolution is random and solutions to the problem of survial can take any number of forms.

"I think that's just a lack of imagination," Losos says, when asked about forms of life similar to those of Earth evolving on alien worlds. "My guess is that an alien world would be very different from what we're familiar with here, even on an Earth-like planet where the physical laws are the same. I think that what we've seen from life here on Earth is that there's a large range of ways in which life can adapt to its environment."

Losos does admit that there may be some similarities as life evolves on alien worlds: it wouldn't be hard to imagine a streamlined body shaped to efficiently move through a fluid environment—just as evolution afforded similar bodies to both sharks and dolphins. But by and large, there's no reason for Earth-like, humanoid life to be the norm throughout the universe.

In Losos's view, this even goes down to the DNA that's at the heart of the heritable traits of natural selection. "Surely there are other ways to transmit heritable information from a parent to an offspring," he says. "It's not even clear that evolution would work the same way as it does here—maybe you would have a system of inheritance that is more like averaging than the system we have. Perhaps competition would be less important and getting along would be more important. It just seems to me that life can go in so many different directions, and to expect that life on another planet would look a lot like ours just seems unlikely."

In place of convergence leading to an inevitable humanoid body shape, Losos points to evolutionary "one-offs"—animals that out-compete and succeed in their own particular niches. These one-offs' solutions to the challenges presented by the environment are not repeated over and over again. For example, the platypus evolved only once, despite the environment for which it was adapted to being repeated in many locations around the planet. Or, closer to home—humans, *Homo sapiens*, evolved in one place, despite other primate populations elsewhere on Earth (including South America, Madagascar, and New Zealand). These locations were not connected to Africa (the evolutionary homeland of our species) and they were primed with early primates, yet no human analogues evolved.

Let's take a test case for the two evolution camps: the Troodon, or, rather more infamously, the Dinosauroid.

Troodon was a small theropod dinosaur that lived near the end of the Cretaceous Period, about 60 to 65 million years ago—late enough to avoid *T. rex*, but unfortunately, just in time for that asteroid to fall in the Yucatán. This small dinosaur had the largest ratio of brain to body weight of any dinosaur of the period, and was as big as a medium-sized chicken. Data suggests that this ratio is something that evolves, time after time, in species after species. As time passes, many species tend to get larger brains.

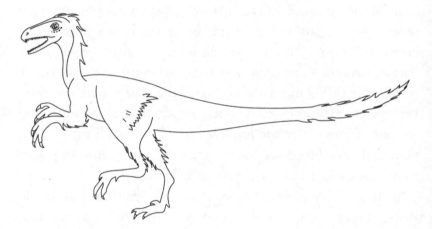

As we've done before, let's speculate on what might have happened if the asteroid hadn't hit. What would things look like today if Troodon had tens of millions of years to evolve? And remember, as far as humans go, the genus *Homo* distinctly evolved only around three million years ago, and modern humans are around 200,000 years evolved from our ancestors. So, tens of millions of years is a good, long time to work the evolutionary mojo.

Canadian paleontologist Dale Russell played this game in the 1980s with Troodon as his subject, and, using then-current evolutionary research and understanding, reasoned that, given millions of years to evolve, Troodon would have turned into a bipedal, humanoid-looking creature with front-facing eyes and a head sitting on top of the body. Think a reptile/lizard person, not unlike the Old Reptile who spoke of Earthlings and his time on Earth in "Interdimensional Cable 2: Tempting Fate." All Russell did was take one aspect of evolution—Troodon's brain size—and follow it through millions of years, and he ended up with lizard people. Think of Lizard Morty attending the Hogwarts analogue in "The Ricklantis Mixup/Tales From the Citadel."

Like any good fan of science fiction, Losos enjoys the concept of lizard people like Troodon, but doesn't see it as an inevitability, here or elsewhere in the universe. In the case of Troodon, Losos admits that brain size most likely would have continued to increase, but that doesn't mean we'd have ended up with lizard people. And add in the knowledge that most (if not all) dinosaurs had feathers of some sort—which wasn't a conclusion when Russell turned Troodon into lizard people—and there's no reason to end up with lizard people.

"If the dinosaurs hadn't gone extinct, would those forms have continued to evolve even bigger and bigger brains?" Losos asks. "Well, maybe, but if they had, there's really no reason to think they would end up looking like humans. They would've been bipedal because those were bipedal animals. It was the loss of the tail that made them upright bipedal animals like us, but if you're more like a chicken or a Troodon, in which case you use your tail to balance your body and large head horizontally above the ground, you very well might keep it. It's not inevitable that if dinosaurs had survived, then their descendants, even the smartest ones, would look like us. Many people have speculated they would've looked like supersized crows that would be very smart, probably using tools, as we see in crows today."

A common illustration of a "modern" Troodon depicts a large chicken or crow-like bird with an enlarged head, thick tail, and arms with feathers and fingers. In at least one depiction, the animal is holding a tiny spear.

Evidence of evolution will be found everywhere there's life in the universe. On that issue, evolutionary biologists agree. Whether the evolutionary pathways that are taken throughout the universe all lead to certain similar points, such as humanoids, or radiate infinitely outward, figuring it out is going to require a lot of portal travel and further study.

DOES EVOLUTION END?

Ultimately, yes. Everything ends. Evolution is an organism's response to changes in its environment, so in order to end evolution, an environment must stop changing. Ultimately, that will happen when the universe winds down and there are no more changes in anything, when the energy of the universe is equally spread throughout space.

Until the universe ends, evolution won't end. It may pause or be directed, but it doesn't end—save for the ending of all life.

Although, the idea of organisms that have "stopped evolving" is one that shows up time and again in science fiction. These creatures are often shown as something that has somehow attained "perfection" in our universe and have been the way they are for millions of years. Often, creatures like this get bored and find novel ways to entertain themselves, often at the expense of "less-evolved" organisms. Something like making an entire planet take part in a musical talent show that pits their best musicians against those of other planets.

While it was never explicitly stated, the Cromulons are similar in appearance and action to aliens who've been said to have reached "perfection" as a species. Whether the Cromulons are or not, the idea of "perfect" organisms who've reached the end of their evolution is a fiction, according to Losos.

"The environment that any organism inhabits is always changing," Losos says. "So even if some life-form could be 'perfect,' the environment would change underneath it and it would have to evolve to remain perfect. Moreover, there is no optimal design, because there's always a trade-off. You can't do everything well; there's always a possibility of getting better at one task at the cost of not doing so well in another task. There's always the possibility of change. It strikes me as unlikely that a form would become perfectly adapted and thus have no need to evolve further."

That brings up the question of the pace or even a pausing of evolution, especially when its participants get too clever for their own good. As Losos explains, humans are—and have been for centuries—tinkering with natural selection. Not just for ourselves but for a wide variety of life on Earth. Given our desires for specific traits in animals, we're playing the role of nature, and we're selecting the kind of organism that survives to pass its genes on to the next generation. In a sense, while natural selection is the strongest force affecting evolution, humans are the strongest force affecting natural selection.

"For evolution by natural selection to occur, there must be some trait that leads you to have more surviving children in the next generation," Losos says. "For the most part, because of birth control, because of cultural norms, and other reasons, that linkage is no longer the case. So, I don't think that for the most part there's much evolution by natural selection occurring in the developed world."

MULTIPLE REALITIES, MULTIPLE EVOLUTIONS

There are a lot of Ricks and Mortys in *Rick and Morty*. A lot.

There are also a lot of universes in *Rick and Morty*.

In his book *Wonderful Life*, evolutionary biologist Stephen Jay Gould came up with one of the most famous thought experiments of evolution, that of rewinding the tape of time back to a point and letting it play again. Gould opined that the way life on Earth played out after that rewind would look nothing like what life looks like on Earth today. Specifically, he mused that we are alive on Earth today because some 505 million years ago, the ancestor of all vertebrates, Pikaia, survived while many thousands of its contemporary life-forms went extinct. If, by chance, Pikaia had gone extinct, we would not be here on Earth, though perhaps some other form of life would. Contingency in a nutshell. Or a fossil bed.

The argument goes that, in terms of life, you never get the same thing twice.

But what about in our beloved TV show? As Rick all too frequently puts it when Morty and Summer annoy him, he has an infinite number of grandchildren, implying that he can always go and grab another. On top of that strong hint, we've seen versions of Summer, Rick, and Jerry who appear, for all intents and purposes, to be identical to their "original" C-137 versions who showed up in the first five episodes—until Rick Cronenberged that Earth, and he and Morty had to move on. There are an infinite number of universes, and therefore an infinite number of Ricks, Mortys, Summers, and everyone else.

Admittedly, Gould wasn't thinking about alternate universes in his thought experiment, and as Losos says, applying his thoughts about "rewinding the tape" to multiple universes may be a little too literal. But the idea can be used as a jumping-off point to discuss evolution on a series of Earths in a way that will keep most evolutionary biologists happy.

First off, this isn't the place to talk about multiple universes. That's actually a few chapters ahead, so I'll just be touching lightly upon it here. Second, if and when physics confirms that there are indeed multiple universes, evolutionary biologists are going to dig in and try to figure out what life would look like if there were multiple Earths.

There are different models of parallel or alternate universes, some where the Earth as we know it would "repeat," and some where it might not. Let's focus on the possible "repeaters."

If alternate universes followed the same physical rules our universe follows, then these parallel universes would have the same number of particles in them, down to the quarks. At a certain time, those particles would all have been in the same place as they were in our universe, and from there, everything would happen in the same way as it did (or does) in our universe. This is Gould's thought experiment in the extreme.

With the same particles in the same place at the same time, following the same progression into the future as they have in other universes, it seems as though you could never end up with anything different—just copy after copy after copy of the same thing. Now we're at the edge of the woods of quantum physics, cosmology, and the philosophical view of a deterministic universe. Follow that path and you'll be wondering about free will and if it's possible that any of us can ever really possess it. That's not where we're going, though.

Losos takes a piece of determinism into account: even if all the particles in alternate universes did start at the same places and times that ours did and followed the same path, uncertainty is baked into everything, including evolution. In other words, as far as Losos is concerned, with multiple universes, you could end up with the same Ricks, different Ricks, or even no Ricks. No big deal.

"If you started two exactly identical worlds, they nonetheless would not experience exactly identical events moving forward, because of the truly random outcome of quantum mechanics," Losos says. "To the extent that's true, that means that over time different mutations might occur, or conditions might be somewhat different, providing the opportunity for life to diverge along different paths and not just those that are completely identical. So, that's sort of the technical argument."

In the broader context, Losos points back to Gould's idea of events that may look like small perturbations having major effects on life thousands of millions of years down the road. "Gould's point really is not that if everything's identical, you'll get the same outcome," he says. "It's more, How resilient is evolution's course against perturbations? And he's really focusing on the idea that something like humans are inevitable. And his response is, no, they're not inevitable. They are the outcome of a particular series of events. And if those events had not occurred, we would have had a very different outcome."

Taking Losos's and Gould's ideas, it may seem difficult to explain the different versions of Rick and Morty and the various worlds that can be found in the series, given that they range from virtually identical to weird versions of one another to not even close to the same. That's just evolution. That's just (again—far more literal than he meant) rewinding Gould's tape and letting it play again. If the alternate universes of Rick and Morty are deterministic, if they all start with the same particles in the same places and move forward from there, there will always be change. Always. That uncertainty is part and parcel of the universe.

Small changes over time would result in small differences between universes. A larger change might result in a divergence that would make a slightly different Rick. Maybe Mexican Rick of the Citadel comes from a universe where the Mexican-American War turned out differently. A series of changes in a rather fortuitous sequence could result in even more bizarre Ricks and Mortys. All the different Ricks and Mortys who've been shown are all just products of their own respective evolutions.

And with a massive number of universes—perhaps an infinite number of alternate universes—the chance that there will be a vast number of similar Ricks and Mortys will be pretty big.

EVOLVING A UNITY

Of all the alien life-forms to look at from an evolutionary point of view, Unity from "Auto Erotic Assimilation" stands out as something unique. To investigate evolution from a single point of view is just like solving a mystery with only the end result and precious few clues presented. From an evolutionary standpoint, again the question is: How in the universe would you end up with something like Unity?

Before we dig in, let's keep in mind that Unity is a genderless alien whose natural form has never been shown. The purple-haired female form that it seemed to favor in the episode was just the alien being through which it was speaking—an avatar. Unity spreads its consciousness to new individuals by means of hosts vomiting into the mouths of others, like an infection that has its own centralized intelligence and consciousness.

Looking at Unity from this standpoint, again the question is: How could evolution end up with something like that? And let's not forget the competitor for Unity's affection, the hive mind named Beta-Seven, a technologically based hive mind. So Unity's not a one-off.

Hive minds like Unity are a staple in science fiction, with literally hundreds of examples spanning movies, television, video games, and comics. Hive minds are characterized by a loss of individuality and identity, resulting in all members working together for a common purpose. As individual creatures each possessing our own individual consciousness, the idea is one of mankind's collective nightmares, and at the same time is presented in speculative fiction as a goal for extreme, autocratic leaders.

To look at the possibility of Unity evolving naturally somewhere in the universe, we can look at examples of something similar on our own planet: colonial animals. Eusociality—a system wherein individuals are arranged in castes and perform a single function while losing the ability to perform other functions—is seen in ants, termites, bees, some crustaceans, possibly naked mole rats, and more throughout the animal kingdom. Of course, while an ant colony is clearly not Unity, the foundation of the hive mind concept is there.

Given the evolutionary distances between the animals that show this behavior, living in colonies has evolved several different times on Earth, suggesting that it's perhaps a typical response to certain environmental factors. Keeping with the ideas of evolution, colonial (or eusocial) animals exist only because they are successful in the larger evolutionary sense. They have a reproductive advantage over those who are not. Eusocial species of animals can defend themselves better against predators and competitors than non-colonial animals, while food distribution among members is, in a primitive sense, "managed," allowing for even the lowest, non-hunting, non-foraging but still important member of the colony to survive.

As for the engine of evolution—reproduction and passing genes on to successive generations—colonial animals such as bees, ants, and termites do have an advantage, even if an individual's genes are not passed on. By virtue of the colony's organization, the genes that get passed on (usually via a queen) are those of your close relatives, many of which you share. Some eusocial animals even have a further advantage in this arena called haplodiploidy, where unfertilized eggs from the queen or other fertile females develop into males of the species, and only fertilized eggs become females. This way, the queen's genes are most often passed on, with mixing via fertilization occurring only occasionally.

Communication between colonial animals can take different forms, from the antenna-based communication of ants to the information-rich "dances" performed by bees to the bizarre touch-based communication of naked mole rats, living underground in near total darkness. The most important method of communication in colonial animals, though, is pheromones—chemical signals released by the organisms. These signals can pass between individuals via touch or through the air, and almost immediately spread throughout the entire colony or all members of a particular caste.

Think of them as nature's wi-fi. While the work of some pheromones is easily seen, such as the trail left by ants to a food source, others can move via diffusion through the air between individuals so quickly that they could be mistaken as a form of telepathy. Studies have revealed pheromones that can, among other behaviors, calm a colony, lay a trail to an important destination, indicate breeding times, and cause the aggregation of individuals to defend against a threat. In the latter case, the signal from the pheromones is strong enough that it can overcome the individual organism's instinct to protect itself from harm. In other words, the organism loses its sense of being an individual and becomes an extension of the larger colony's defense.

Let's bring the discussion back to Unity. Again, it's important to remember that, in the episode, we don't see Unity. We only see

Unity's hosts. As for what Unity actually "is" . . . who knows? The evidence suggests that Unity is most likely a colonial organism that is contained in the vomit that it uses to spread itself among hosts and that can thus hijack a host's systems for its own uses. Each individual performs a certain task that is beneficial to the larger "organism" or colony. The other individuals are under the direct control of Unity's constant communication—as we see, distraction can cause individuals to break from their main consciousness. Whether that form of communication is a nudge that gets the individual moving on the path Unity needs or Unity is in direct control of every individual all the time isn't quite clear. That communication may be telepathic or via a very efficient pheromone system—the two methods would be similar in appearance and function.

Taken into consideration with Earth-based examples, the idea of a Unity-style hive mind evolving in the universe is something that can still be classified as fiction. But, given its appearance in nature, maybe not far-flung fiction.

EVOLUTION IN FROOPYLAND, COURTESY OF TOMMY LIPNIP

Evolution is something that happens in response to changes in the environment. Take away those changes, and, in theory, evolution shouldn't occur. Organisms shouldn't change, and everything should stay in a state of homeostasis. True, it would be an artificial place—say, a place that was created by a father for his daughter to be completely safe and that was full of imaginary friends (with their own "imaginary DNA") where she could enjoy her childhood. A father might think of it as a masterwork and a tangible symbol of his love for his daughter (or a place to put her to protect the neighborhood from her sociopathic behaviors), although one day, she may come to see it as a "glorified chicken coop."

That's Froopyland—or at least it was Froopyland from "The ABCs of Beth."

Rick designed Froopyland, with its procedurally generated clouds, rivers that were rainbows, breathable water in which you couldn't drown, bouncy ground, and harmless creatures, to be a completely safe place. To do so, he created a stable, unchanging environment. Froopyland's stability is something that is completely artificial, and it's also a death sentence for evolution. The environment never changes—there are no pressures or natural selection on the creatures—so therefore, the organisms created by Rick never change.

Or should never have changed, or rather would never have changed, if Beth hadn't left Tommy inside Froopyland after he fell (or was pushed) into the Honey Swamp.

As Rick and Beth found, the introduction of Tommy caused changes.

Tommy's permanent addition to Froopyland changed the environment. An artificial change to an artificial environment, to be sure, but still, it was changed. Tommy needed to eat, and so he made his own food, introducing genetic diversity in the process.

As twisted as this cycle was, Tommy was putting a selective pressure on the population of Froopyland's creatures. He would mate with the creatures, they'd (quickly) have offspring, and Tommy would eat the ones that were tasty. The less-than-tasty ones he would let live. The population shifted and changed.

Like a creature shuddering to life, evolution began in Froopyland.

Tommy's less-than-tasty offspring needed to eat as well. As he apparently upped his production of offspring, evolution really got roaring along. A population of not-very-tasty children grew, along with a menagerie of other creatures.

When Beth returned to Froopyland a dozen or so years after the Honey Swamp incident, the place had changed. Now there were predatory birdlike Froopy creatures, a population of simple (apparently not very tasty) beings, and probably more. The collection of Froopyland creatures seen in "The ABCs of Beth" is the result of natural selection over the years. Dangerous creatures that Tommy could not subdue (such as the bird with talons that attacked Rick) would be left to populate Froopyland. Docile creatures would be used as breeding stock. The tastier offspring would be eaten, the not-so-tasty allowed to grow into maturity.

Given that the new Froopyland creatures were hybrids—a mix between human DNA and Rick's "procedural carbon"—a safe assumption would be that they would be sterile. And yet, the carnivorous bird-creature had babies in its nest. But Tommy was a coward. There's no way he would have used a vicious bird as breeding stock. That bird reproduced naturally. Froopyland creatures are able to reproduce, and each successive generation introduces more variability to the population. Had we seen more of Froopyland, we certainly would've seen more exotic creatures that themselves would have been putting pressure on the environment and other creatures living there, and causing more changes.

And that diverse mix of creatures should be expected. As shown in the episode, the gestation period for the creatures is extremely short; therefore, a dozen or so years in Froopyland could have created thousands of generations of creatures. With the changes in the environment that Tommy introduced, thousands of generations is more than enough time for evolution and the creation of the new species that were seen. After all, we're less than eight thousand generations away from the very first *Homo sapiens*.

Given more time, Froopyland would have looked nothing like Rick's original design, perhaps even approaching a microverse level of organization, thanks to the slow changes of evolution over time.

No matter the place, as long as there's life, there will be evolution.

CHAPTER 5

Black Holes, Wormholes, and Exoplanets

★ ★ ★ ★ ★

In *Rick and Morty*, there are a lot of planets out there. There is also a means of traveling from place to place that's most likely based on artificial wormholes, though they also exist naturally. By looking at exoplanets and wormholes, we can grasp two different methods by which science works: observation and inference. Both give us a chance to show just how clever we—that is, *Homo sapiens*—are.

When we look at astrophysical or cosmological phenomena, we can see amazing wonders; but if we want to get there, scale is what stands in our way. Distances to other stars or planets outside our own solar system make travel there impossible. Yet we know these things exist. Or should exist. Or could exist.

Let's start with the things we can "see": exoplanets.

DISCOVERY WITH OBSERVATION: EXOPLANETS

Rick and Morty makes good use of the fact that the galaxy, the universe, and the multiverse are all filled with planets. In science fiction, this isn't anything new. Every show, movie, comic book, or video game set in space has its share of alien worlds—usually full of alien life and peril.

Even in our universe, the idea of exoplanets—planets that exist outside our own solar system—has been around for centuries: Italian philosopher Giordano Bruno theorized their existance in the 1500s, Isaac Newton posited the idea of other solar systems in his *Principia*, and many other astronomers and scientists throughout the years have looked to the skies and realized that there were probably more planets out there around those stars that were so very far away.

But we didn't have even a hint of evidence of exoplanets orbiting other stars until 1988, when Canadian researchers Bruce Campbell, Gordon Walker, and Stephenson Yang, making observations at the extreme limits of their instruments, tentatively suggested that a planet might be orbiting the star Gamma Cephei, forty-five light-years away. Subsequent observations in the years following went back and forth on the is it/isn't it game, until the existence of the planet was solidly confirmed in 2003. In the thirty or so years since that first detection, we've gone from none to almost four thousand, and that number is primed to explode as new technology and tools are set to come online in the next few years. The planets that have been discovered range from rocky chunks of nothing to planets orbiting two stars to waterworlds to Earth-like, massive giants that make Jupiter look tiny and many more.

Most exoplanets are so far away that we can't see them directly. So how do we know they're there? How we find them is almost as interesting as the exoplanets themselves, and it's a great example of how science is used to find new things. In almost all cases, exoplanets were found by following a simple "if/then" framework: "if distant stars have planets orbiting them, then . . ." with the "then" being used in the methods by which the planets are found.

In short, if you know what should be happening with these hypothesized planets orbiting other stars, you should be able to see them when you look for them.

Let's look at the major methods by which exoplanets are discovered.

TRANSIT

You know that scene in a horror movie in which there's a doorway at the end of a long, dark hallway, and suddenly there's a shadow in front of it, but then it's gone. Was it the killer? Where did they go? Instead of killers and hallways, imagine a star light-years away. As you look at the star, occasionally something apparently passes in front of it, causing the amount of light to dip just a little. That thing is most likely a planet, transiting in front of its star.

That's how most exoplanets have been discovered: the light from distant stars decreases just a little as the planet passes between its star and our telescopes. The amount of dimming is far too small for the human eye to perceive, so other instruments are used, such as the (recently retired) Kepler Space Telescope and the Transiting Exoplanet Survey Satellite (TESS).

When a planet transits in front of its star, scientists can determine all sorts of information about it: the planet's size, its atmosphere, even its composition and temperature. That's how we know that there are exoplanets with methane, oxygen, and water vapor in their atmospheres, even though we've never laid eyes on them.

The transit method isn't perfect, though, and some planets, orbiting at just the right position, could be invisible to our instruments, while other signals that seem to indicate the presence of a planet may be false positives. Luckily, there are other methods that can confirm what transit data suggests. Still, nearly three thousand exoplanets have been discovered with this method.

RADIAL VELOCITY

A star-planet system is not a series of perfectly nested circles (or even ellipses) placidly moving around one another. Get that idea out of your head now. Stars and planets have dynamic relationships, wherein the planet pulls on the star while the star pulls on the planet. The star is, of course, many, many times larger than the planet, and as a result, the planet orbits the star. But the planet's effects on the star cause it to wobble away from the center of gravity of the planet-star system—just a little. Small planet, small wobble. Big planet, big wobble. More than one planet, complicated wobble.

From a vantage point near Earth, this wobbling star has regular and predictable movements that take it ever so slightly closer to Earth and then farther away again. This causes a change in the light that can be observed coming from the star, due to the Doppler shift. Everyone has experienced a Doppler shift, or, as it's better known outside astronomy, the Doppler Effect. Energy travels as waves, and the length of those waves can change depending upon how the wave source is moving. An ambulance's siren sounds higher-pitched as it approaches you, and then lower-pitched as it passes and moves away from you. As it approaches, the wavelengths of the sound are shortened (the source is catching up with the waves), resulting in a higher pitch, and as it moves away, the wavelengths are elongated, resulting in a lower pitch. The passengers in the ambulance don't notice any of this. Their frame of reference for the siren is different from yours.

As the star moves away from Earth, the light waves coming from it are stretched a little, and their wavelength is increased. Red light has the longest wavelengths, so as something is moving away from Earth, it's said to be "redshifted." If the star is moving toward Earth again, even minutely, the wavelengths of light are squeezed just a little. Blue light has shorter wavelengths, and as such, objects moving toward Earth are said to be "blueshifted."

Looking at stars with planets, or whole planetary systems, we see that the light modulates through the color spectrum, thanks to the wobble caused by the tugs and pushes of the bodies that cause the star to jiggle toward Earth and then away, over and over, periodically. When the data is analyzed with other instruments, the amount of color shift can indicate the planet's mass (bigger, denser planets have more pull than smaller, less dense ones, which may not move their star by an amount detectable by Earth-bound tools), and that data can be combined with other methods of detection to build a fuller, more complete picture of the exoplanet.

At no time is an exoplanet ever seen by radial velocity alone. While this method has resulted in some exoplanet discoveries, it is more often used in conjunction with other methods, such as transit, to confirm suspected exoplanets.

GRAVITATIONAL MICROLENSING

Thanks to Albert Einstein, we know that massive objects, like stars, can warp space-time. As a result of this warping, light itself gets bent and distorted as it travels near these objects. Under the right conditions, and with the right objects, light can bend in a fashion that can be observed and measured using instruments on Earth.

If a star with its attendant exoplanet has enough mass, it can cause the light coming from an object behind it to be focused periodically, resulting in a brighter-dimmer-brighter pattern as it travels. Spotting a microlensing event such as this is rare, but they have been observed, especially when large portions of the sky are viewed for long periods of time, a task made easier by the invention of the automated telescope. From the microlensing event, the mass of the object or objects causing the phenomenon can be calculated.

DIRECT IMAGING

Sometimes, exoplanet hunters can just get lucky. If a planet (or planets) is orbiting at just the right distance from its star, it can be discovered the old-fashioned way: by looking at it. Well, first scientists have to block the light from the planet's star, the same way you would hold your hand up to cover the sun so you can more easily spot something outside on a sunny day. The "hand" in this case is either a coronagraph—a disc inside a telescope that can be positioned over a star—or a starshade, which is a separate spacecraft positioned near a space telescope that can be maneuvered to block starlight from various stars, allowing astronomers to observe the planets in their orbit.

This method is best for large, hot planets whose infrared radiation can be seen with specialized telescopes on Earth, but it also works with planets orbiting small brown dwarfs (which aren't full-fledged stars). It's even been used to spot rogue planets—planets that wander through space without a star.

When the James Webb Space Telescope is launched in the early 2020s, it will be able to spot exoplanets (and a lot more) via direct imaging.

ASTROMETRY

The oldest of the methods used by exoplanet hunters, astrometry dates to the nineteenth century, when astronomers kept careful records of the positions of stars in the night sky using photographic plates. Like radial velocity, astrometry looks for the slight positional changes of a star over time caused by the orbit of its planet or planets. This method's early success was the classification of binary star systems in which two stars orbited around each other, their positions shifting slightly as each responded to the other's gravitational pull.

These methods, along with a handful of others, have resulted in the discovery of nearly four thousand exoplanets, all by observation of the object or the effects of the object on another, more visible object.

But what if the object can't be spotted and may or may not have effects on anything nearby? Would such an object even be considered to be real?

A HYPOTHESIZED DISCOVERY WITHOUT OBSERVATION: WORMHOLES

In the world of *Rick and Morty*, wormholes are real. Rick and Morty know that—it's how Rick and Jerry returned to Earth after the events of "The Whirly Dirly Conspiracy." A wormhole was (almost) used by Fart to get to his universe in "Mortynight Run." And Rick's Portal Gun is likely a wormhole generator that can get him from place to place in an instant.

In our world, too, wormholes are real. Or they should be. Or they might be. Or we might be able to make one someday.

Open any popular astrophysics or relativity text and you'll more than likely find a pretty standard illustration showing space-time looking like a piece of paper bent over on itself, with a wormhole connecting two otherwise incredibly distant points. Wormholes are

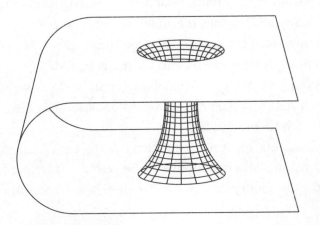

shortcuts through hyperspace, or the area outside our universe's regular "space," connecting one region to another region far away. Prominent astrophysicists talk about wormholes in documentaries, write about them in books, and comment on them in public talks. And they were discovered in math.

Wormholes are an example of a "discovery" in science made without direct, observable evidence. They're hypothetical, but no one thinks that one is going to show up anytime soon; we're not even likely to spot one by gazing out into the stars. While hypothetical, wormholes do have solid footing on which to stand: Einstein's theory of general relativity. Wormholes are not specifically prohibited by the theory, so that means they can exist. Somehow . . . somewhere. Again, maybe.

While that argument may sound like a lame justification from a teenager ("Well, you didn't specifically say that I couldn't go to that party, so I thought that meant I could . . ."), in physics, it's actually an accepted way of saying something probably exists. Einstein's theory of general relativity demonstrated that the laws of physics are the same for all non-accelerating observers, and also that the speed of light in a vacuum is the same for everyone, everywhere, no matter the speed at which you're traveling. Wormholes aren't even mentioned.

From that relatively short explanation (and the equations—there's always math), Einstein and other physicists were able to better define and explain the universe, while predicting aspects of the universe that were, at the time the theory was published in November 1915, impossible to prove. As a model, general relativity is our best means of explaining the universe as we observe it, while predicting phenomena yet to be observed. That's what good models do in science.

Some aspects of general relativity were tested during Einstein's life, and others continue to be tested today; predictions based on the theory have yet to be shown to be wrong. Six of the most important predictions from general relativity that were shown to be true include: the prediction that explained why Mercury orbited the sun in the weird manner it did (which defied Isaac Newton's explanation of gravity); the prediction that large masses, such as the sun, would bend light (a phenomenon known as gravitational lensing, as mentioned earlier); the prediction that light should become more red as it moves past a massive object; the prediction that a large mass would fractionally slow light's speed; the prediction that the laws of nature are the same everywhere, and always have been; and the prediction that all masses distort space-time, and large masses pull space-time along with them as they spin.

On the surface, the predictions based on general relatively can look insane. But in test after test, they've all been proven true, even though Einstein never did any of the experiments. He just figured out the relationships (and thus the equations). To this day, the theory keeps getting tested, and it keeps passing each test. For example, the detection of gravitational waves that won a team of astronomers the 2017 Nobel Prize in Physics was predicted by Einstein's theory. Again, in mid-2018, a group of astronomers showed that general relativity holds true on massive, interstellar scales. Using the gravitational lensing characteristics of a distant galaxy to infer how much mass the galaxy contained, they showed results that were just what general relativity predicted, on a scale much, much larger than anything previously seen.

The collected ideas, then, are physicists' guidelines for what can and what can't exist, cosmologically and astrophysically speaking, in our universe. General relativity sets up the conditions that explain how our universe runs, and if your idea can work within the guidelines of the theory, then the idea—or object, or cosmic phenomenon—could be out there somewhere.

Case in point: black holes. General relativity predicts that an object of large-enough mass could distort space-time to a point where nothing would be able to escape its pull—not even light. The idea is a natural part of the evolution of stars: large stars, spinning slowly enough, will eventually implode and form black holes, which will be much smaller than the original stars and will have such massive gravity that even light cannot escape from them. No light coming out from them, no light being reflected off them, thus they are "black."

The modern idea of black holes was first introduced by Karl Schwarzschild in 1916, but it wasn't until the late 1950s and early '60s when more work on the idea was done, and when they were given their iconic, ominous name, by physicist John Wheeler. While astronomers had been able to point to the effects of suspected black holes (a cluster of stars orbiting something that can't be seen, or X-rays or radio waves coming out of the centers of galaxies, for example), it took us until 2019 to capture an image of a black hole, a supermassive one named M87*. All the initial ideas about black holes were founded on general relativity's tenets, not on direct observation. It took decades for astronomers to piece together the picture of what and where black holes are, and even longer to be able to capture an image of one.

It's a similar story for wormholes. These astrophysical phenomena were "discovered" mathematically in 1916, when Ludwig Flamm worked out a solution to Einstein's field equation that allowed for the existence of such things. A few years later, working with physicist Nathan Rosen, Einstein himself studied the ideas and proposed that there may be "bridges" that cut through space-time.

These passageways were what we call "wormholes," more properly called "Einstein-Rosen bridges," and were thought to have two mouths and a "throat" that connects them. Others worked with the concept over the decades, further clarifying what these things would be like, even though they couldn't directly observe them.

If you're keeping track, Rick knows all of this. He's seen more than his share of general-relativity-allowed phenomena—

including a fair amount of stuff that we can't even imagine. The point in tying our world's experience with general relativity to the world of *Rick and Morty* is that wormholes exist in both universes. In the world of *Rick and Morty*, they're just there, a function of the storytelling by the writers. In our world, they're a function of Einstein's theory of general relativity, and are elusive.

Once researchers proved the existence of wormholes mathematically, research meandered for a while, as if not quite sure what to make of this oddity. Again, the "research" was all done within the boundaries of general relativity and Einstein's other calculations; none of it was done by direct observation because, even today, we have no tools to observe wormholes, and only a possible idea of where to look. But based on decades of research, we can discuss what we expect wormholes might be like when we do find them.

"Wormholes apparently do exist, but the trouble is the ones that we know exist are inside black holes," says Richard Matzner, a cosmologist at the University of Texas at Austin. "Although a black hole is something that sits there eternally, the inside of a black hole is dynamic, and can create wormholes that open pathways to somewhere else. But this wormhole exists and then collapses back and pinches off in less time than it would take you to get through it, so it would not be a pleasant experience to try to use it."

The black-hole-connected wormholes described by Matzner are of the Einstein-Rosen variety. A singularity in the middle of a black hole may be able to almost instantaneously reach out to another singularity through hyperspace, form a channel, and then pinch off again, letting the pieces of space-time "bounce back" to their original locations.

But not every naturally occurring black hole produces a wormhole, and not every wormhole is connected to a black hole—at least, in theory. Another idea suggests that wormholes may be quite small—small enough to the point of uselessness for humans to travel through—but perhaps an integral part of how our universe works. "At very, very small scales, the universe is really not foamy but sort of wormy, like it's sort of riddled with wormholes," Matzner says. "This would let things communicate over distances they ordinarily wouldn't [be able to], but it's still an incredibly microscopic scale. 10^{-33} meters is the scale that this is happening at, and that's very, very small."

Just an aside about that scale of 10^{-33}: that's as big as a . . . well, okay, there's nothing at all that we regularly run into at that scale. Neutrinos, infinitesimally small particles, are around a yoctometer in diameter, and that's 10^{-24} meters. We don't even have a prefix for 10^{-33} meters. At that scale, you're talking about hypothesized bits of the universe, like quantum foam and the actual "strings" of string theory. Something that's 10^{-33} meters, is very, very small—maybe on the order of the smallest that there is.

Matzner's explanation doesn't describe wormholes as they're seen in science fiction. The prevailing theory of wormholes is that they're very transient and, more than likely, submicroscopic—like a minuscule bubble that forms but immediately pops. Admittedly, there's very little "hole" to a wormhole using this description, and there's no way to use it for traveling as seen on *Rick and Morty*.

So, general relativity and Einstein predict that wormholes can exist, but admittedly, a wormhole that would be useful for human purposes needs to be large and to stay open long enough for things to get through.

The answer to wormholes staying open came in the form of parallel lines of research and thought experiments: one as a product of black hole research by Stephen Hawking and George F. R. Ellis in the 1970s, and the other from a simple request for help from one scientist to another for a story he was writing.

In that latter example, the scientist-turning-science-fiction-author was Carl Sagan, and the story was *Contact* (1985). In the novel (later turned into a movie starring Jodie Foster), Sagan's protagonist, Ellie Arroway, traveled from Earth to Vega, a star twenty-six light-years away. Sagan's original story had Arroway dive into a black hole near Earth, travel through hyperspace, and emerge via a black hole near Vega an hour later. Something told Sagan he needed some help on the method of transport, so he called his colleague (and expert on relativity and black holes) Kip Thorne.

Thorne immediately realized that Sagan was in trouble with using a black hole as transport, since there's way too much radiation and energy inside one for anything to make it out the other side (leaving aside the insane gravitational forces). This got Thorne thinking about wormholes as a solution to Arroway's interstellar travel problem. He knew about the probable existence of wormholes, but their life-span and propensity to evaporate and destroy anything traveling through them made them just as unsuitable as black holes. Thorne needed to come up with a way to prop a wormhole open.

Thorne's solution: so-called exotic material (or exotic matter). This unknown material would have unique antigravity qualities that would push the walls of the wormhole apart, as well as a negative energy density that could result in a stable wormhole. The idea Thorne presented to Sagan was to drop the idea of travel via black holes and to use spherical wormholes instead, threading them with exotic matter like scaffolding to hold them open. Sagan used the idea in Contact, and, intrigued, Thorne dug further into the idea of wormholes and exotic matter. Contact, by the way, was riffed on in one of the flashbacks of "Morty's Mind Blowers," when Rick and Morty had to trick NASA scientists into trading places with them in an alien's menagerie via a ship built with instructions sent by Rick. And if you were wondering, it was this introduction in Contact by Sagan via Thorne that thrust the notion of wormholes into the public, science fictional, and cultural consciousness.

So wormholes are permitted by Einstein's general theory of relativity, and much of the serious, scientific work done to figure out how they could work for transportation came as part of making a science fiction story work. Again, though, there was no direct observation.

Thorne's suggestion of exotic matter to keep the wormhole open required him to characterize the properties of such material, and in doing so, he was able to match many of the characteristics of his proposed exotic matter with real conditions that exist in the universe. In short, the broad strokes of Thorne's suggestions about stabilizing a wormhole are possible in theory, although it should be said that the exotic matter conditions that would be necessary for a hypothetical stable wormhole may preclude any travel through it, as the exotic matter would interact with the traveler as well as the wormhole. But still, Thorne's approach to the problem—asking big, radical questions that were almost science fictional in nature—led him to discoveries and insights that perhaps would not have been possible otherwise—discoveries that were made in minds, in scribbled notes, and after a multitude of worked equations.

The method by which Thorne reached his conclusions about exotic matter came as a result of what he calls "Sagan-type questions." That is, basically, what could an infinitely advanced civilization do if the only limits were the laws of physics? These are thought experiments: grandiose ideas triggered by bold, what-if-style questions. As a sidenote, Thorne's love of big questions and exploring big ideas through story led him to provide the scientific basis for and serve as executive producer on the movie *Interstellar*. He was also on the team that won the 2017 Nobel Prize for the discovery of gravitational waves.

Even though his work was fundamental in explaining how wormholes would work and could be used, Thorne and other astrophysicists and cosmologists remain pessimistic on the idea of naturally occurring wormholes. To bolster this point, as far as we understand the universe and general relativity, there's nothing out there that could naturally become a wormhole as part of its evolution or life cycle. The formation of black holes, both small and large, can be predicted from the age and size of stars, but there's nothing astronomers can point to that will, one day, become a wormhole.

As mentioned earlier, wormholes may exist on the smallest scale, in matter referred to as quantum foam. In this idea, the entire universe would be absolutely riddled with wormholes that are vanishingly small and continuously blinking in and out of existence, all nestled in quantum foam. Some researchers even speculate that that wormhole-riddled quantum foam is somehow the connection between the theories of quantum gravity and relativity: entangled objects that are far apart are able to "communicate" with each other at vast distances because, by this theory, they are actually very close, via a network of submicroscopic wormholes in the quantum foam.

All of this leaves us in a weird spot. Einstein's theory of general relativity allows for wormholes to exist, and there's an idea that they might exist, very briefly, inside black holes, but it's possible that they do not exist naturally. But it's a big universe.

"We haven't seen everything even in our own corner of the universe," Matzner says. "For the past fifty years we have had astronomical and cosmological surprises every decade. So there is still a chance that an astrophysical phenomenon that does produce macroscopic wormholes will be discovered. It's still unlikely, I think, but possible."

Which keeps the door slightly open to the idea that if we (or a super-intelligent alien species . . . or a guy named Rick Sanchez) could figure out a way to make a traversable wormhole, then the universe would—or should, at least—allow for that particular perturbation. More on making wormholes with Portal Guns later . . .

CHAPTER 6

Portal Gun

★ ★ ✦ ★ ★

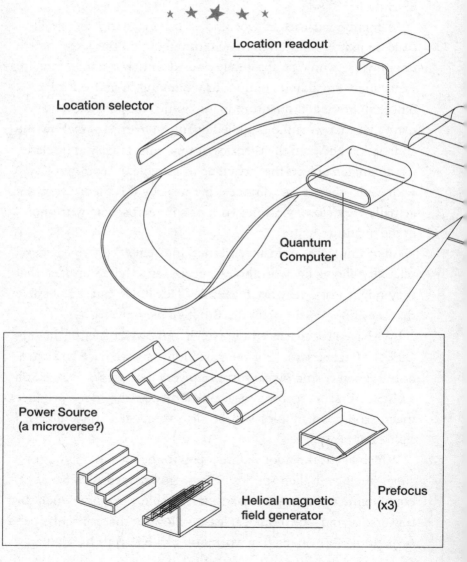

Location readout

Location selector

Quantum
Computer

Power Source
(a microverse?)

Helical magnetic
field generator

Prefocus
(x3)

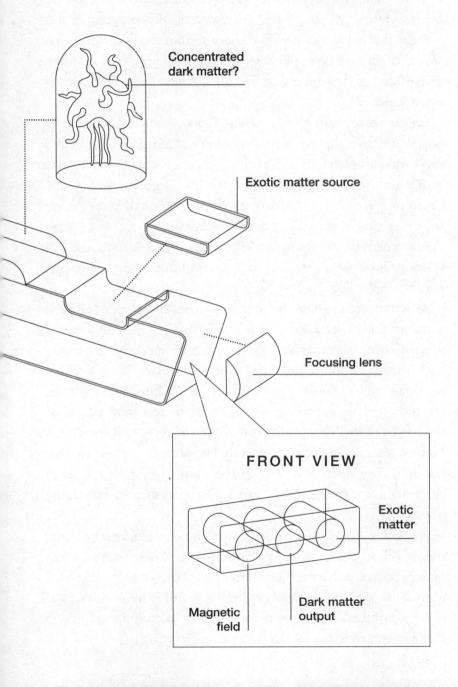

Concentrated
dark matter?

Exotic matter source

Focusing lens

FRONT VIEW

Exotic
matter

Magnetic
field

Dark matter
output

Rick's Portal Gun is many things: it's a story element in "The Rickshank Rickdemption." It's a deus ex machina, allowing for last-minute escapes from otherwise deadly encounters. It's a cheat that allows interstellar and interdimensional travel that would otherwise be impossible. It helps move the plot at times, and, of course, it's the reason for the Galactic Federation's interest in Rick.

Unless there's some other scientific explanation for what the portal gun does that the writers of *Rick and Morty* haven't shared yet, it's a safe bet that Rick's Portal Gun opens a wormhole. It looks simple on the screen—Rick shoots the gun, a green or blue hole opens in space, Rick and Morty step through, and then they're on the other side of the galaxy or in another universe; that's a standard element of science fiction. But when we think about the real science that could allow that to happen in our world, things get really complicated. Fast.

As we covered earlier, the first type to be "discovered" and the simplest version of a wormhole is one that is spherical, created when two singularities in different parts of space reach toward each other. Think of one singularity as a black hole, sucking things in, and the other as an opposite, or "white" hole, spitting matter out. They are connected for an impossibly short period of time via a tunnel that pinches off almost as soon as it is formed. Later research showed that wormholes can take various shapes, but still, they open and close in less than an eyeblink. They're certainly not traversable, at least as we currently understand the physics.

For Rick's Portal Gun to open a wormhole to a known location, it needs three things: a way to make a wormhole traversable, an energy source, and a means of finding the correct destination. It's a small list, and Rick's already figured it all out in his universe, but in our universe, those three make travel by wormhole something that's relegated to the far, far future . . . if ever.

Oh, and a few more things before we dive into making wormholes that we can travel through: Rick's Portal Gun does follow some internal logic. Portals with blue borders are most often used for intergalactic (i.e., "local") travel, while those with green event horizons seem to open onto alternate universes. Also, portal technology, and the intelligence to make it, is rare, regulated, and highly sought after in the universe of *Rick and Morty*. Consider the trouble the Galactic Federation went through to get the formula for interdimensional travel out of Rick's mind in "The Rickshank Rickdemption," as well as the lengths to which Rick went to protect that information—the construction of an entire false set of memories. The Federation's intergalactic portals are large and bulky compared to Rick's gun, which can open them—and interdimensional portals—anywhere. And let's not even get started on the US government's portal technology, which Rick described as "ghetto-ass," among other things.

The reasoning for the Federation's regulation of portal technology is easily understood: portal technology could be world-, galaxy-, and universe-changing tech. For example, militaries could sweep in and conquer their enemies from all directions almost instantaneously. No location would be secure from infiltration. Populations would start to shift en masse, since you'd no longer need to live anywhere near your job. Portal tech would also cause massive disruptions in transportation, government, privacy, and a million other aspects of life as we know it. And you were thinking only about being able to instantly go to a beach in the Caribbean after work, weren't you?

Also, as we dig into the Portal Gun and what may make one work, we should probably discuss Rick's spacecraft as well. As the series creators have mentioned in interviews, Rick's Portal Gun plugs into a docking station in the dashboard of the ship, allowing the ship and its passengers to travel through both intergalactic and interdimensional portals.

The question of who originally created portal technology will be an issue of debate among fans for years. Was it Rick C-137, and the information he managed to leak to the Ricks of other dimensions? Or was it another Rick, and Rick C-137 got the idea from other-Rick (as his constructed memories seem to suggest)? Or is it just something that all Ricks in all dimensions will eventually create, and by virtue of their traveling between dimensions, be accepted into the fraternity of Ricks? That's a question for another time. For now, let's worry about Portal Guns.

One last caveat: I don't distinguish that much between intergalactic and interdimensional portals in what follows. The larger concept is the same, whether your destination is six hundred light-years or seven alternate realities away. You need to exert enough energy to punch through space-time and establish a traversable pathway. The hypothetical method will be roughly the same, with the chief difference being energy use, but we'll get to that later.

HOLD THAT WORMHOLE OPEN

Let's say you can create a wormhole. As best we know how, that means a singularity with a black hole at either end. This means you can also somehow create black holes and keep them under control so they don't eat the living room. They're just there for you to walk through so you can get to the party at the other end.

But that tube between black holes wants to close in on itself something fierce. You're going to need to find a way to keep it open.

As explained earlier, it was CalTech physicist Kip Thorne (and later his students Mike Morris and Ulvi Yurtsever) who first played with the idea of making wormholes traversable by employing Einstein's field equations to find the conditions that would allow for a wormhole to remain open. What they found was that the wormhole would have to be threaded with a material that has negative energy. They called this material "exotic matter," due to its behaving the opposite way from normal matter; it would have negative mass as well, since Einstein showed that mass and energy are equivalent with $E=mc^2$—that is, if E (energy) is negative in the equation, m (mass) would have to be negative as well to make the equation work, since c (the speed of light) is a constant and has a positive value.

In other words, the material near the mouths of the wormhole would have to have a mass or energy density that is less than empty space. As a result, the exotic matter would have a repulsive gravitational effect and would effectively prop the wormhole (which wants to pinch off and close) open for as long as the exotic matter stays near the respective mouths. Keep the mouths open and the tunnel portion, or the "throat" as it's called by astrophysicists, will stay there. Probably.

Easy, right?

Not exactly.

In the years since Thorne and his students suggested exotic matter as a way to hold wormholes open, the topic has remained active in research. The use of exotic matter to hold wormholes open isn't limited to spherical wormholes, either; any wormhole of any shape or orientation is traversable only if it is threaded by exotic matter. Both wormholes themselves and the existence and use of exotic matter see several papers published in academic physics journals each year. This isn't weird science-fiction stuff limited to fringe scientists. It may sound that way, thanks to the theoretical physics, but it's all grounded in solid theory.

But, as Richard Matzner explains, wading into the idea of exotic matter gets somewhat trippy, somewhat quickly. For example, we know of one type of exotic matter: dark matter. But while dark matter may be exotic matter, it may not be the exotic matter we're looking for to hold wormholes open.

"Dark matter is a substance that pervades the whole universe and has a very small effect on short scales," Matzner says. "But as you get farther and farther away from, say, a center of something, like a galaxy, the dark matter pushes you harder and harder, with more and more acceleration, away from that center. So I'd say that's a probable example of this exotic matter. It's just not strong enough for practical use."

Not to mention that once you have this wormhole threaded with exotic matter, you must walk (or fly) through it. You'd be exposed to the exotic matter, and its effects on humans, depending upon who you ask, range from mostly harmless to every atom in your body rushing away from every other atom at nearly the speed of light, give or take a few thousand meters per second. That does kind of take the whole "traversable" out of traversable wormholes.

Although it should be noted that physicist Matt Visser, who's picked up the wormhole baton from Thorne, has suggested that nonspherical wormholes could be constructed with a thin layer of exotic matter, and, as a result, could reduce or eliminate the exposure of the traveler to the exotic matter. Visser's suggested shapes include a cubical wormhole with the struts built out of exotic matter, polyhedral wormholes, and other nonsymmetrical shapes that would prevent exposure.

The best bet is a constructed wormhole, at least as far as we can put the theoretical pieces together. It would need to have an artificial-looking structure to it—a cube or a polyhedron—in order to ensure both stability and the safety of the wormhole's travelers.

But finding that exotic matter is still a problem. Even if dark matter is the exotic matter we need, we have yet to be able to observe it directly in the universe, let alone think about containing it. And it's probably not green and fluidlike and something that can be stored in an inverted-jar-type thing on your wormhole-creating device, either. So, you've probably got to make it. And that's a trick too.

Matzner explains, "If you have two uncharged but conducting metal plates and you put them very close together, say, nanometers apart, they can actually show a repulsive force between them. This is because you're actually removing some of the energy density of the vacuum. So, the vacuum, which has nothing in it, apparently has zero energy too, but when you bring the plates together, it produces this strange repulsive energy."

This weird effect is called the Casimir effect, and its causes can be explained by quantum field theory, zero-point energy, and how both relate to the vacuum (at the quantum level, vacuums have a very complex structure). With the Casimir effect, both attraction and repulsion have been demonstrated in the laboratory, but there's a catch: the Casimir effect has only been shown to have effects in the nanometer range—that's billionths of a meter. This is not entirely helpful for creating something that a human-sized being can walk through, let alone drive a spaceship through. A repulsive force, however, as long as it's grounded in quantum field theory, is maybe a foundation for the type of exotic matter we'd be looking for.

But why create when you could just find and manipulate?

As discussed earlier, space-time may be, at its smallest, most fundamental level, foamy in nature, riddled with a near-infinite number of wormholes that constantly appear and disappear. While we're talking about (from our understanding) implausible technology, we should take into account the idea of grabbing a submicroscopic wormhole, somehow expanding it up to macroscopic size, and then threading it with exotic matter. A shortcut to figuring all of this out, Matzner says, is kind of simple. Just find a wormhole somewhere in the universe. We're pretty good at figuring out how things are done if we can look at how the universe did it first. "It took us a while to figure out that there were nuclear reactions in the sun, but we did, and we learned from that and moved forward from there. If we're ever going to figure out how to do this, it'll be because we found some astrophysical object that is doing it. Once you see something can be done, then you can figure out how to do it."

And finally, the most recent research on wormholes proposes that a species of wormhole may exist that does not need to be threaded with exotic matter. As in, it would be traversable and would remain open on its own. The negative energy in this case would come from a quantum connection between the pair of black holes that serve as the wormhole's mouths. Under the right conditions, something entering one black hole could exit the other unharmed. The physicists behind this research, Ping Gao, Daniel Jafferis, and Aron Wall, point out that this version of wormholes shares a lot in common with the idea of quantum teleportation, and at a certain level, one is virtually indistinguishable from the other.

The Gao-Jafferis-Wall version of a wormhole is an offshoot of a larger argument in physics that is summed up with the equation "ER = EPR"—that is, Einstein-Rosen (ER) bridges, or wormholes, are equivalent somehow to entangled quantum particles, or Einstein-Poldolsky-Rosen (EPR) pairs.

Entanglement is something that can occur, essentially, between two particles. When a measurement of one is made, the other immediately demonstrates a corresponding state—they react to being observed. For example, measure the spin of two entangled particles. If one is seen to be spinning clockwise, the other, no matter how far away, will be spinning counterclockwise. But the creepy thing is that until the actual measurement, neither has a specific state of spin—it's as if they're somehow connected ("entangled") and can communicate over any distance instantly. This has been demonstrated conclusively and repeatedly in laboratory research. The larger ER=EPR argument is the most recent attempt to connect the general relativity of Einstein with the quantum realm of entanglement, in order to understand gravity at the quantum level (which, by the way, we don't).

Einstein called entanglement—this connection that appears to defy the rules of relativity—"spooky action at a distance." Yeah, no kidding.

We'll come back to the idea of entanglement in a little bit.

THIS IS GOING TO TAKE A LOT OF JUICE

The existence and creation of wormholes is theoretically possible.

In this case, "theoretically" means that the energy needs of the wormhole construction aren't a roadblock (in energy terms), and that you'll be able to find a whole lot of energy to make this thing happen once the science is figured out. But it will take a *lot* of energy, which must come from somewhere.

Rick's Portal Gun can run out of energy—it happened in the pilot episode and a few more times over the course of the first three seasons. And it needs energy to run. Probably a lot.

As Matzner says, there are more than a few energy problems in creating a wormhole. Even just manipulating the conditions around an existing wormhole would require a huge amount of energy to create and thread the exotic matter. And that's if you weren't starting from scratch.

You've probably heard a physicist somewhere, when asked how much energy would be required to accomplish a particular "big science" question, say it would take a sun's worth of energy. You probably—not entirely incorrectly—thought that's just the physicist's way of saying, "Get out of here, kid, you're bothering me."

In this case, it's probably more true than not. As best we can understand it, starting from scratch and punching a hole through one area of space into another would take an insane amount of energy—somewhere on the order of Planck energy, which is around 10^{19} billion electron volts. That's in the neighborhood of Big Bang–level energies. The Large Hadron Collider (LHC) outside Geneva, Switzerland, can produce around 14 trillion electron volts (or 1.4×10^{10}), which isn't going to be tearing holes in space-time anytime soon.

Other theories for coaxing wormholes out of existing super-massive black holes—for instance, with a helical magnetic field and a proposed type of dark matter called axions—would also require similar amounts of energy that can be found near supermassive black holes, as particles are accelerated to near-light speed.

With Planck-ish levels of energy concentrated in one spot, we could possibly encourage the formation of a wormhole out of the foam at the bottom floor of space-time. To do either, though, the amount of energy required would be on a similarly massive order of magnitude. But there's a problem here: it's such a large amount of energy that, as best physics can predict, space and time would start to break down, and the laws of physics would start to erode as well. We just don't know what would happen.

The amount of energy in the universe at the point of the Big Bang might be enough to pop open a passageway between two points in our universe, or even between our universe and another. We just need a new Big Bang. The thing is, the idea of creating "baby universes" in a lab isn't all that weird; it's been around since the late 1980s. Cosmologist Alex Vilenkin explained a mechanism in which a universe could have started when there was no time, space, or matter. According to Vilenkin, a bubble of space could just appear and then begin to inflate spontaneously to a universe-sized scale. And this method could someday possibly be controlled or replicated on purpose, creating new universes.

Other physicists have taken this idea further and have described means by which, through a more powerful particle accelerator than the LHC and a monopole (a hypothetical particle), a new universe could be grown "next door," connected to our universe by a wormhole. Growing a universe—it's becoming more of a technological issue than a totally theoretical or hypothetical one.

Rick has a miniverse that serves as the battery for his car, so we know he has no qualms about using the power output of a planet to give himself cold-cranking amps. Why not steal the energy from a just-started baby universe (destroying it in the process) to open a portal to get from point A to point B?

On the scale of Rick-ness, sucking all the energy out of a new universe to get pizza from a different dimension is at least an 8 out of 10.

HOW DO WE KNOW WHERE WE'RE GOING?

The US government's portal technology in *Rick and Morty* answers the question of arriving at your correct destination in a rather low-tech way—there has to be a transmitter "mouth" for you to enter, and a receiver "mouth" where you exit. This is the same level of technology used by the Galactic Federation, and pretty much a standard view of wormhole technology. It also could be seen as synching with Thorne's ideas: your transmitter and receiver could be made of exotic matter, or could even create the exotic matter on site, thereby stabilizing your wormhole. Then, just use a touch of computer/GPS guidance to point one toward the other and you're good to go. Still very much science fiction, though, based on the exotic matter and energy issues noted above.

Getting somewhere that you haven't been before—and therefore haven't set up a receiver—would be trickier, but there are theories as to how you might do it. Remember entanglement? Einstein's "spooky action at a distance" has already been invoked in the pairing of some forms of black holes or even quarks that are connected via a wormhole. The opening of one mouth of a wormhole may be, in some manner, connected to another mouth via entanglement. Nudge the other mouth a little and you may be able to get it to where you need to be.

A slightly different option, as Matzner explains, would be that, once inside the wormhole, it may be possible to perform some level of navigation that would result in the "steering" of the wormhole to the precise destination. Of course, with a powerful-enough computer, this "steering" could be done nearly instantaneously, to make it appear as if your path was always set and never wavered once you got into the wormhole.

At this point of trying to navigate your wormhole, we've reached a place where even theory fizzles. We came from a place as near to certainty as general relativity allows (that, technically, wormhole creation may not be impossible) and left that certainty to talk about energy needs. But now, when it comes to guiding a wormhole to a specific destination, ask any physicist, and most will fall back on Thorne's original thoughts in his advice to Sagan, and his own ideas in *Interstellar*: the creation of traversable wormholes might be possible by an ultra-advanced civilization, one whose knowledge of science and ability to manipulate the laws of physics far outstrips ours.

An ultra-advanced civilization . . . or maybe just an incredibly smart scientist who stumbled onto the most powerful technology in the universe and uses it for adventures with his grandkids.

CHAPTER 7

Multiverse Theory

★ ★ ✦ ★ ★

D on't go feeling like you're special or anything. There's a good chance you're just one of an infinite number of yous. And, honestly, some of them are more interesting than this version is.

How do we even know the multiverse exists if we've never seen any universe other than our own? While it looks, in all likelihood, like the multiverse does exist, there have been physicists from the very emergence of this theory who disagree, claiming that the "evidence" is nothing but mathematical oddities among the calculations that will be smoothed out as we learn more.

Two of the leading advocates of multiverse theory are physicists Brian Greene and Max Tegmark. Each has come up with his own multiverse models, Greene with nine and Tegmark with four. Both scientists are on solid theoretical footing when it comes to thinking about the existence of a multiverse, based on predictions of accepted theories whose other predictions have been shown to be true.

As physics grows broader in scope, the possibilities for proving at least one of the theories about a multiverse inch ever closer to reality. As such, while there are still multiverse skeptics, a large number of physicists now believe at least one version of what Greene and Tegmark have proposed, or something very close to it.

Before we go on, let's cover some terminology:

* Universe (interchangeable with Cosmos): where we live. Everything we can see and measure is in here with us. Anything that we can't see or measure is not part of our universe.
* Multiverse: a collection, either finite or infinite, of universes.
* Alternate timeline/dimension/reality: terminology mainly used in science fiction to thread a particular needle and set some parameters for what's being talked about. Most physicists prefer "alternate universe."

HOW FAR CAN WE SEE?
THE OBSERVABLE UNIVERSE

Fair warning from the start: the idea of the observable universe can mess with you. Back in school, you may have been asked, "How big is the observable universe?" as the final question on a test.

In this scenario, you think about it and realize it's pretty simple. Like a person sitting on an island in the middle of a large body of water who can see only to where the sky meets the ocean, an observer on Earth can see only a given distance in any direction, to what is referred to as the cosmic horizon. We should, therefore, be able to see things that started emitting light when the universe started. That should be the boundary of the observable universe.

Now that you've got more of an idea, you can start developing that thought. You remember that distance equals time multiplied by speed, which is a simple calculation. The Big Bang happened around 13.8 billion years ago, and since light travels at 3.0×10^8 meters per second, it travels 9.5 billion kilometers per year (5.9 trillion miles). Multiply the time by the speed and you get 1.3×10^{20} km, about 81 billion trillion miles, which equals about

13.8 billion light-years. We can see 13.8 billion light-years in any direction, so Earth sits at the center of a sphere that has a radius of 13.8 billion light-years. Simple. Put your name at the top of the paper, turn it in, and sit and look around the classroom with a smug look on your face.

But your classmates are still working.

As you feel the prickly sweat start to rise up on your skin, you wonder if you could have been wrong with your very logical line of reasoning. As your teacher will undoubtedly explain to you, your answer is indeed wrong, because you forgot a few things.

The actual time when visible light started moving through the universe is a little shy of 13.8 billion years. For about 380,000 years after the actual Big Bang, the universe was: 1) smaller, and 2) hotter. The universe was so hot that the newly formed particles were moving too quickly to pair up. That meant that protons and electrons were zooming around but not able to lock in and hold on to one another. This happens all day, every day near us: matter is heated to such a point that atoms lose their electrons (called ionization) and run around naked in a new state of matter. The state of matter is called plasma, and it's what our sun, and all stars, are made from.

Plasma is opaque to photons, which make up visible light. Photons bounced around between the charged particles of the primordial universe. They just couldn't get anywhere. For the first 380,000 or so years, the universe was a big, glowing, universe-sized blob of plasma.

Then something amazing happened: recombination.

The universe was expanding during its plasma phase, and as it expanded, it cooled. As the particles cooled, they moved slower, ultimately allowing protons to hook up with electrons, forming the first and most abundant element in the universe: hydrogen. The plasma dissipated, and light was free to spread throughout the universe.

While the difference in the time won't change the calculated size of the universe that much, it is important to note that visible light wasn't shining from the very start. It took a little while.

The reason your calculation is so far off is, put simply, expansion.

Space has been expanding since the Big Bang's first push. Everything is still spreading out. Some parts are moving away from one another faster than the speed of light, which on its surface sounds wrong. But Einstein's special theory of relativity enforces the universe's speed limit, saying that no matter can move through space faster than the speed of light, but the theory is silent about the expansion of space relative to us. Space itself can expand faster than the speed of light.

Once we begin to consider expansion, things start to become clearer. An object that emitted light, say, 10 billion years ago is not 10 billion light-years away from us. Since it emitted the light that we're seeing now, that star has continued to move, making its actual distance much farther than the 10 billion light-years.

The change in light's "start date" in the universe and the expansion of space each play a role in making the observable universe larger than your original calculation. But there's one more piece to the puzzle. As data from the Hubble Space Telescope and other instruments has shown, the expansion is accelerating. It's getting faster and faster.

Putting all of this together, our best estimate for the size of the observable universe is a sphere with a radius of 46.5 billion light-years, with us at the center. No, Earth is not at the center of the universe—it's at the center of the *observable* universe, relative to us, because we're the observers. If the Smiths were interested in looking at the sky while they lived on Dwarf Terrace-9 in "The Wedding Squanchers," their observable universe would have been slightly different from that of Earth. Given that Dwarf Terrace-9 was in the Milky Way, the difference would have been minimal because of the distances involved in measuring the observable universe.

The observable universe is exactly that: it's what we can see. There's no good cosmological reason to think that the universe doesn't continue beyond our cosmic horizon. As for what may lie past what we can see . . . that's the first version of a multiverse.

THE UNIVERSES NEXT DOOR . . .

The first two versions of a multiverse are fairly easy to grasp, once the idea of the observable universe has settled in. For these, no jumping dimensions or weirder "realities" are needed.

If we leave Earth and take a trip past the edge of our cosmic horizon, there's no reason to think that we wouldn't just enter a new area of the larger universe. And the universe just goes on and on and on for an infinite distance.

The most common speculation is that these new areas, or "universes," have the same physical laws and constants as ours but are just so far away that there's no way our friends back on Earth could ever encounter them. The "new" area of the larger universe would be, for all intents and purposes, a "different" universe from the one we just left and knew as home. Yes, there would be overlap of areas when considering the observers, but let's focus on a more or less central point from which observations are made.

This is a pretty simple way of thinking of a multiverse. Astrophysicist Brian Greene compares this way of viewing a multiverse to a patchwork quilt or a Quilted Multiverse. Barring hitherto unknown technology, observers in the center of one patch would never know of, or be able to interact with, observers in the center of an adjacent patch, let alone those four or nine patches away. And this quilt goes on forever. Literally forever: it's infinite. And if it's infinite, there are bound to be repeats.

Another type of multiverse model assumes that the stretching of the cosmos—called inflation—is eternal. As a concept, this goes back to the Big Bang and what happened immediately after. In 1980, theoretical physicist Alan Guth theorized that just after the "event" of the Big Bang, the universe moved into an inflationary period that lasted from 10^{-36} seconds to 10^{-32} seconds. In other words, inflation of the universe started a trillionth of a trillionth of a trillionth of a second after the Big Bang itself, and lasted slightly less time than that.

During the inflationary period, the universe went from a sub-sub-sub microscopic point to, well, "universe-sized." In essence, the universe grew exponentially from nothing into everything in a period of time our poor, addled monkey brains would see as "instant."

Guth's theory of inflation has been rigorously shown to work by a number of very precise observations made over the decades since it was introduced, and is now considered by most physicists to be the best model for what happened just after the Big Bang. Which is a good thing, since it explains many of the features and measurements we see in our universe, such as the Cosmological Principle, which states that the average density of galaxies throughout the universe appears to be equal. Inflation also fits nicely with the observations of the universe being homogeneous and isotopic, or "smooth," with all its matter more or less evenly distributed, while showing an evenness of temperature throughout.

But even with something as seemingly simple as inflation, things can get weird. Inflation doesn't have to stop everywhere all at once. Physicists refer to its progress as being chaotic. Where inflation does stop or pause, you get a Big Bang—a "hot" Big Bang—like the one that gave rise to our observable universe. But beyond that, inflation may not have stopped, leaving a region of empty space. After that region, inflation may have stopped again, kick-starting another universe's Big Bang, but continued elsewhere. The end result: multiple universes separated by empty, inflating space. These universes would be unknowable and unreachable to one another. Instead of the patchwork quilt of the previous example, think of a tabletop with marbles scattered on it in random locations or a field of bubbles. Each marble or bubble is a universe. The heavy lifting on the development of this model was done by Stephen Hawking and Jim Hartle.

These ideas fit into Greene's Inflationary Multiverse model. In this model, inflation (with its universe-creating hiccups) goes on forever, continuously creating new "bubble universes." This process also creates an infinite number of universes. To simplify one of Greene's favorite metaphors for this model of the multiverse, think of reality like a block of Swiss cheese. The bubbles (which look like holes when you slice it) in the cheese are the multiple universes of reality, while the cheese itself is space. Moreover, the block of cheese is constantly expanding, and the number of bubbles inside it is increasing as well.

As I mentioned at the very start of this chapter, this idea of the multiverse is a prediction based on theory, or, rather, on larger frameworks of physics: inflation and quantum field theory, both of which make other predictions that have been proven true. As more observations and research appear to sharpen the conclusion that inflation is eternal, then a multiverse made of bubble universes is just a consequence of that conclusion.

There are three other important considerations while we're thinking of these types of multiverse models:

First, unlike the patchwork quilt model, there's no reason to think that these bubble universes will share the same physical laws and constants. Some may have more (or fewer) dimensions than our universe (ours has four, including time). Some may have a strong, pervasive magnetic field running throughout. Some may have weaker gravity or other different values for the fundamental forces of physics. Some may be vibrant and (we assume) able to support life like ours, while others may be dark, sterile, universe-sized volumes of scattered quarks and electrons that never formed complex matter.

Folding the idea of the inflationary multiverse in with string theory, which allows for different conditions and topographies of space, Greene proposes an even more complex version of a multiverse called the Landscape Multiverse.

The second consideration for bubbles and quilts is that inside a bubble could be a quilt universe. From inside the bubble, the universe would appear infinite with no edges, while from outside the bubble, the universe would be finite, encased by the bubble's "walls." While counterintuitive, both views work: a seemingly infinite universe (as seen from the inside) contained in a bubble, a universe that's a functional, patchwork quilt multiverse; a multiverse inside a multiverse. Our universe, with its presumed patchwork multiverse that exists beyond what's observable, could be inside a bubble.

Finally, the third consideration is somewhat depressing for those of us without Portal Guns. The patchwork quilt and bubble universe models share something in common: universes are separated by space. Not dimensions, not realities—just a space. Or, more simply, distance.

This may seem pretty hopeful in the sense that if you were to travel far enough and for long enough, you could find another universe, and maybe find doppelgängers of Earth and everyone on it. Or, at best, maybe a fresh Earth for a do-over. But in reality, there's no possible way to reach other universes by travel. Our universe is expanding at an accelerating rate, and as best we understand it, inflation hasn't stopped. Everything is moving away from everything else, and on top of that, space itself (the block of cheese from earlier) is getting bigger faster than you could travel. Until Portal Gun technology shows up, get used to living in this universe.

ETERNAL + INFINITE = INFINITE RICKS + INFINITE MORTYS

Let's go back to one of the observations that fits both with our universe and the idea of inflation: the isotropic nature of the universe—basically, that it's the same in all directions. Through measurements taken by the Wilkinson Microwave Anisotropy Probe (WMAP) of the cosmic microwave background radiation fluctuations (the changes in the energy "echo" of the Big Bang), NASA researchers were able to measure the density of the universe (how much stuff is in it, volume-wise) with great accuracy. No matter where you look in the universe, the average volume is made up of 4.6 percent atoms (matter as we know it), 24 percent dark matter, and 71.4 percent dark energy. That's the same everywhere. Everywhere.

Current thought about inflation gives no reason to think that this critical density would change in any other bubble universe, though the laws governing the matter may. In short, the same amount of stuff is inside each universe.

Let's focus on matter. We experience matter as atoms that are made up of protons, neutrons, and electrons. Protons and neutrons are made from specific combinations of two types of quarks, which, along with electrons, combine with each other in any number of ways to make the matter we're made from and with which we interact. Think of the particles like clothes. Let's say you have ten T-shirts and seven different sets of pants. Taken together, you have 10 shirts that can be worn with any of the various pants, so 10 x 7 = 70 different outfits. If you had to be seen at 70 different functions, you could do so without ever repeating an outfit. Every style would be unique. But if there are 140 functions, you're going to repeat outfits. In other words, the possible outcomes exceed the possible combinations. The number of repeated outfits increases as the number of functions increases (obviously).

What if there are an infinite number of functions? You would repeat those outfits a lot of times. An infinite number of times, in fact, because there are an infinite number of functions. Likewise, if there are an infinite number of functions, there are only so many ways you can act or behave. As with your outfits, you'll repeat your actions at the functions an infinite number of times.

Let's bring things back to *Rick and Morty*. The characters throw around the word "infinite" a lot. The concept is always floating in Rick's mind when Morty or Summer are in peril or annoying him. Let the one die, and go and find another. There are apparently an infinite number of "regular" Ricks and Mortys, not to mention the variations seen at the Transdimensional Citadel of Ricks: cowboy Rick and Morty, robot Rick and Morty, aquatic Rick and Morty, cyclops Rick and Morty, Eric Stoltz Mask Morty, and all the rest.

Infinite functions, finite outfits, finite actions and behaviors. Infinite universes, finite particles, finite actions and behaviors. There are going to be some repeats. An infinite number of repeats. That's at the core of the idea of "infinite."

Rick isn't being hyperbolic when he tells Morty that he has an "infinite number of sisters" at the start of "Rickmancing the Stone." He does. Likewise, there are an infinite number of variations of Summer, Morty, and Rick out there, from the nearly identical to the ridiculous, as we've seen.

But some of the possible parallel universes of the two types explained so far might be inaccessible or hostile to life, yielding no Ricks, Mortys, Summers, or anyone else, only piles of quarks and electrons that, had conditions been different, would have been assembled into Rick, Morty, Summer, and the rest of the universe.

So let's talk about the Central Finite Curve. There are a lot of Ricks out there to keep organized, and the Central Finite Curve is the measuring stick by which it's done in *Rick and Morty*.

When Rick protested his treatment by the Council of Ricks in

"Close Rick-Counters of the Rick Kind," the Council's Chair said, "Of all the Ricks in the Central Finite Curve, you're the malcontent, the rogue." Simple Rick, whose memories were used to flavor Simple Rick's Simple Wafers Wafer Cookie, was located sixty iterations off the Central Finite Curve.

Think of the Central Finite Curve as a way for the Council to keep track of the Ricks—and for the series's creative staff to do so as well. In a Q&A about the Pocket Mortys game, series co-creator Justin Roiland said, "When we talk about the multiverse, we use the word 'infinite' a lot; and when we think about 'infinite,' our brains melt. There is a Central Finite Curve of accessible realities, and they get weirder and weirder as you get to the edges of that curve, but past that curve is just garbage and inhospitable static realities."

Roiland's "inhospitable static realities" easily match the idea that some bubble universes' physical laws and constants would be different and, as a result, would not be fine-tuned for life. To use the earlier example of a universe where fundamental particles never joined to form complex matter: there's no Rick there.

Speaking of infinity melting brains . . . a selection of infinite number of realities would be infinite, not finite. Five percent of infinity is infinity. So taking the Council at its word, the Central Finite Curve is most likely just a collection of realities and Ricks that have been discovered and dealt with, or, at the very least, are considered interesting enough to be counted. Perhaps the Ricks who were tasked with finding and categorizing all the Ricks worked until they felt they'd found enough and called it a day.

The easiest way to think of the Central Finite Curve would be as a frequency or normal distribution, or, more accurately, a Gaussian function. Infinite realities would be on the x-axis, with the measure of "Rickness" (presumably as determined by the Council) on the y-axis. The "Rickness" of each reality's Rick would be plotted against its reality, forming a bell-shaped curve with a peak in the middle and tails off to either side that lead to less and less "Rickness." Since Rick C-137 has expressed many times that he is the "Rickest Rick," he may be the Rick at the center of the Central Finite Curve.

Doofus Rick of Dimension J19-Zeta-7, seen in "Close Rick-Counters of the Rick Kind," is probably a little farther off the curve.

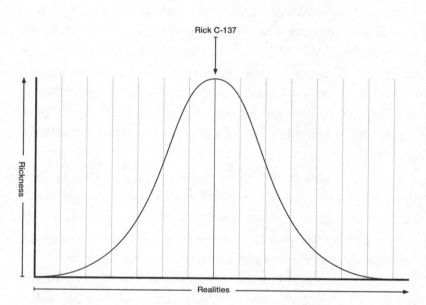

OTHER MULTIVERSE FLAVORS

Another view of multiple universes by Greene sees them as something akin to slices of bread, stacked in a larger framework. This is the brane theory of multiverses, and starts with the assumption, from string theory, that reality in fact has nine (or more) spatial dimensions. This model works by taking a bunch of universes, each with their own four dimensions (three of space, one of time), and embedding them in higher-dimensional space—a region with more than the four dimensions we're used to—which is called the "bulk." Each universe of the larger multiverse is a "brane" ("membrane" abbreviated) within this larger structure.

Instead of a patchwork quilt or marbles, the multiple universes of brane theory exist like slices in the aforementioned loaf of bread. Matter is stuck to its own brane, although Greene suggests that gravity may leak, tentacle-like, between branes. While controversial, the idea of leaky gravity is used by some physicists to explain that gravity is such a weak force in our universe and apparently has no particle to mediate its effects (like the way electromagnetic force has the photon) because gravity may not be from our universe.

This idea of a multiverse is perhaps the most common (or second most; see the many worlds theory below) version of a multiverse that people think of when they hear the term. This is the multiverse model of "stepping between worlds" shown countless times, from the Chronicles of Narnia to DC Comics' Earth 2 to the "Mirror, Mirror" universe of *Star Trek*.

What makes brane theory particularly compelling is that there may be a way to run experiments that could possibly suggest its existence. As proposed by Greene, crashing particles in the Large Hadron Collider may eject debris off our brane and into the bulk. Given that the ejected debris would take the energy of the collision with them, if the energy before the collision does not match the energy after the collision in just the right way, this could indicate that the energy left our brane—that it's left our universe and gone into the higher-dimensional space in which the branes "float."

Another of Greene's multiverse theories grew as an offshoot from brane theory. In the cyclic multiverse model, branes oscillate back and forth and could occasionally collide with one another, and the energy of the collision would be enough to kick off a localized Big Bang, which would create new universes that are contained in their own branes. These cyclic universes would have their own limited life span (perhaps a trillion years), beginning and ending as the parent branes "shudder," snapping back and forth against each other over and over. Conveniently, the idea of a cyclic multiverse quiets the question of "How did the universe begin?" The branes have always been and always will be oscillating.

And, of course, within the universe of each brane, that universe would be infinite, meaning that you've got a patchwork quilt multiverse thing going on, and there's no reason to think that each brane wouldn't be undergoing eternal inflation. Add an infinite number of branes on top of that, and you get multiverses on top of multiverses on top of multiverses.

Feeling insignificant yet? Just wait.

OTHER MULTIVERSE FLAVORS: THE MANY WORLDS THEORY

While the patchwork quilt, bubble, and brane multiverses may take a little cosmology and understanding of how the universe works in order to get going, the many worlds idea of the multiverse is much more intuitive, thanks, in large part, to its use (some may argue overuse) in science fiction and pop culture—even before it was formally developed by Hugh Everett III in 1957.

To get into the many worlds idea of a multiverse, let's close our eyes and think about Charles Dickens's *A Christmas Carol*. The climax of the story comes when the Ghost of Christmas Future shows Scrooge his own grave—that of a forgotten, petty man. Motivated by the sight of his headstone (and other events of the night), Scrooge changes his ways and vows to be a better man. And he is. Happy ending. Merry Christmas.

But in the many worlds hypothesis, he is and he also isn't. In the published version of the story, Scrooge changes his ways. But also, in other worlds, Scrooge told the spirit to take a hike, continued being an ass, and did end up in that grave. One decision point, two possibilities; both choices were made, and thus two realities exist as different universes—Good Scrooge universe and Bad Scrooge universe. And so on, for every decision Scrooge makes thereafter.

Everett's theory can be seen as an answer to the Schrödinger's Cat thought experiment: a cat is in a closed box with poison that may or may not be released. The mechanism controlling the poison's release is connected to a radioactive element, the decay of which is a completely random process, wholly out of the hands of any observer. Until the box is opened and the cat's condition is measured by an external observer, the cat is stuck in a weird dead/alive limbo. Once the box is opened, the two possibilities collapse upon each other and the cat is either alive or dead, and the universe continues on, with either a living cat or a dead one.

Not content to settle for an either/or solution, Everett basically asked, "Why not both?" The universe doesn't choose. Both things happened: the cat was alive and the cat was dead. If you're looking at a dead cat in a box, then that's your universe. Meanwhile, in a universe that just split from yours, the cat is happily meowing in the box, as cats sometimes do. Every action causes a split. Think about the decisions you made today—then add all the ones you made yesterday, and so on. And then add in roughly the same number for all 7.5 billion people on Earth. The number of universes seems to run up to infinity, although there are some suggestions that it doesn't exactly reach infinity . . . just a 1 with 500 zeroes behind it. Which is still a lot.

We're used to seeing this approach to a multiverse in fiction, such as in the movie *Sliding Doors*, the "Turn Left" episode of *Doctor Who*'s Tennant years, and Rick's explanation after the Cronenberging of Earth-137 in "Rick Potion #9," in which the epidemic grossly mutated nearly all normal life on Earth and Rick and Morty simply jumped into a similar reality where that wasn't the case anymore and where those versions of them had just died at a similar time.

One clarification, though: thinking that it's the conscious decisions of humans that split reality is an oversimplification made to help wrap brains around the larger idea. As the hypothesis works, each occasion of quantum activity will result in a new universe.

Unlike the previous versions of multiverses, many worlds has some solid (albeit weird) experimental proof to back it up. One proof: shoot an electron at a barrier with two slits, and it will act like a wave of energy and go through both, if there is no observation at the point where the electron goes through the slits. Look at the screen at which the electrons are being shot, and you'll see a pattern characteristic of waves. Repeat the experiment, but this time observe the process at the point where the electron passes through the slits, and it will pass through one slit or the other as if it is a distinct particle. This experiment and others like it have been repeated thousands of times, with the same results.

Also, the different universes are not some unfathomable distance apart, or embedded in the larger bulk. All the bifurcated universes of the many worlds are believed to exist with us, in something called Hilbert space, which has infinite dimensions. Travel between the variations is impossible, as far as we know—unless you have a Portal Gun.

This version of a multiverse is also what was broadly played with in the second-season episode "A Rickle in Time" when, due to time being stopped for six months, Morty and Summer accidentally split their timeline. The split takes the entire Smith house out of space and time and into its own pocket reality, a reality populated by floating cats. As I mentioned earlier when talking about time, there was a nice nod here to the basis of the many worlds hypothesis—Rick assumes that the cats are Schrödinger's Cats, all stuck in limbo, just as they are.

Through the episode, further uncertainty splits each timeline until there are ultimately sixty-four different realities, representing splits that happened at different points of uncertainty.

EVEN WEIRDER MULTIVERSAL FLAVORS

Greene's full list of science-based ideas for a multiverse includes two more ideas: simulated and holographic.

Greene's theory of simulated multiverses is relatively easy to consider. Each universe would be a simulation filled with self-aware bits of code. Like other multiverse ideas, there could be an infinite number of such universes, from the near-identical to the wildly different. Travel between simulated universes could be something as simple as an external intelligence taking a Fortnite character and putting them in *World of Warcraft*, or, if the universe's inhabitants have developed their own technology, they may figure out a way to get from universe to universe, à la Wreck-It Ralph.

And, finally . . . the idea of our universe being a hologram first came about in the 1990s and, to date, has as much evidence supporting it as there is supporting the theory of inflation. The holographic universe is a product of information theory and suggests that everything that makes up the reality we experience as three-dimensional is contained in a two-dimensional surface. Think of it like the hologram on a credit or bank card: the object in the picture appears to be three-dimensional, but from your perspective, it's clearly two-dimensional. But if the 3-D object— say, a bird—was self-aware, it would feel as if it were in a 3-D environment.

This suggests that our reality, our "real" physical reality, is actually located (or, to use a more apt word, encoded) on a lower-dimensional boundary, and what we experience in the hologram has three full dimensions. The surface that we mirror with precision would be far away, such as at the edge of the universe. At its core, the holographic multiverse is a simpler multiverse, made up of just two "universes": the real and the reflection. Travel to the edge of the universe and you'd eventually meet your "real" self traveling toward you.

A FINAL MULTIVERSE FLAVOR:
THE ULTIMATE MULTIVERSE

Think of this as ordering "one with everything."

In this view of the multiverse, everything—all types of universes—exist. A universe of literal nothing. A universe with an absurd number of spatial dimensions. A universe of cheese. A universe based only on Newtonian mechanics, inhabited, Greene posits, by solid billiard balls.

At the core of the differences between the universes in this model is math. Our universe has its own specific set of mathematical laws that it follows. This proposed type of "ultimate" multiverse comes as the result of pondering why that's the case. If this one universe that we know, with its specific math exists, then every universe must exist, because there is an endless set of math.

Admittedly, this version of a multiverse is a touch more philosophical than cosmological.

A FINAL VOICE OF REASON?

Just after Stephen Hawking's death in 2018, one of his final papers was published in the *Journal of High Energy Physics*. Written with physicist Thomas Hertog, the paper, titled "A Smooth Exit from Eternal Inflation," looked to tame the idea of an infinite multiverse. Hawking admitted that he had never been a fan of the infinite multiverse, largely due to the problem of "sciencing" it. If there are infinite multiverses, each with a random set of physical laws and constants, there's no way to test it. If something cannot be tested, it's on shaky ground as far as science is concerned. It may be a lovely idea, but, as Hawking and Hertog said, it's just not science.

Specifically, they were looking at the idea of infinite and chaotic "bubble" universes, each with its own possible physical constants and physics. Utilizing views of holography, gravitational waves, and string theory, Hawking and Hertog's new theory cuts the number of possible universes down to a finite number, and existence becomes "more constrained, more uniform," Hertog says. The new theory makes the idea of a multiverse more predictable and testable. In reaching this conclusion, the scientists also suggest that there can only be universes with laws of physics identical to our own, as well as that our universe is, in fact, just typical. We're not that special—but the assumption of ours being an average universe means that the observations we make in our universe can most likely apply to other universes.

While the new theory about multiverses does seem to go against some of the other leading ideas and models, there may be a way to prove it—with direct evidence from the Big Bang, probably in the form of ancient gravitational waves with massive wavelengths. The hunt for such waves is currently underway, and the results may one day indicate that Hawking and Hertog's theory is correct, and our universe isn't all that special.

"OKAY, THIS ALL SOUNDS LIKE CRAP . . ."

. . . is hopefully not what you're saying right now, but you might be. And you wouldn't be alone in the larger science community. While some physicists accept the ideas laid out here in some form, some take a more conservative approach to the larger ideas, or leave them for others to refine and prove. And there are those who outright reject them as nothing more than mathematical aberrations, fanciful imaginings, or, in the case of the ulitmate multiverse, worse.

But here's the thing, as Greene, Tegmark, and many others explain: these theories aren't just made-up science reflecting an absence of evidence. Inflation and other theories make claims and predictions that then go on to be tested—and some of those claims are shown to be true. That's how science works. You start with a theory, and if its predictions are true, it is adopted as an explanation for how the universe works. The theory only lives as long as observation and predictions agree with its claims, but it begins as a suggestion.

The idea of a multiverse (perhaps excepting the simulated multiverse) is something that agrees with at least one accepted theory of how the universe works. Observations and predictions based on these accepted theories have been shown to be true over and over. Following through, all Greene and Tegmark are suggesting is that if the theories have stood up and been shown to be true via observation and prediction, then the idea that we live in a multiverse is one that must be considered with equal possibility, and even validity, as others.

However, testing and making observations of multiverse predictions are virtually impossible at the present time. This is the issue that has some in the scientific community wanting to keep multiverses on the "philosophy" side of the science/philosophy border. Multiverses may be a terrific collection of ideas, but if there's no way to observe them or to test their characteristics, well, science has pretty strict rules about that.

All of this being said, think of big ideas of the past. We used to think that the Earth was the center of the solar system. We used to think that our sun was the center of galaxy. We used to think that our galaxy was the only one. We used to think that the universe was static and unchanging. Each one of those claims is missing its punchline: "until someone proved otherwise, and changed our view of the universe."

That may just happen again.

CHAPTER 8

Why Dimension C-137?

★ ★ ✦ ★ ★

If you wanted to advertise to the rest of the universe that you were smart and in on the secrets of the universe, you'd broadcast the number 137, the fine-structure constant. It's probably a safe bet that the creators of *Rick and Morty* were in on the same idea when they gave the Rick and Morty that we first meet the designation "C-137." If you understand the nod to 137, you know this is a show for you.

The fine-structure constant was first found in 1916 by physicist Arnold Sommerfeld (it's also occasionally called Sommerfeld's constant), who found the value when he was describing the motion of electrons around a hydrogen nucleus. The value also accounts for the splitting of spectral lines in the light produced by an excited hydrogen atom. It's these closely spaced groups of lines that are referred to as the "fine structure." Since its introduction, the fine-structure constant has shown up over and over in physics and quantum electrodynamics (QED), and manifests in the physical world in a myriad of ways. This recurrence of the fine-structure constant in various places has had a profound effect on many physicists over the years: it's weirded them out.

Nobel Prize–winning physicist Wolfgang Pauli was obsessed

with the number and its appearance in quantum mechanics, relativity, and electromagnetism. Somewhat morbidly, Pauli joked that his first question to the devil after he died would be for an explanation of the fine-structure constant. Pauli even sought out help from famed psychiatrist Carl Jung to help him gain some insight on the significance of the number.

Perhaps appropriately (or, again, weirdly), Pauli died in room 137 of the Red Cross hospital in Zurich in December 1958. The room number wasn't unknown to Pauli as he lay on his deathbed. Reportedly, the coincidence unnerved him.

Another Nobel Prize winner, physicist Richard Feynman, poked and prodded at the fine-structure constant as well, calling it "a magic number that comes to us with no understanding by man." He also suggested that all physicists should hang a sign in their office with "137" on it, to remind them of all that they do not know and keep them humble. In pointing to more bizarreness surrounding the number 137, Feynman was convinced that, due to electrons having to orbit the nucleus of atoms with more than 137 protons faster than the speed of light, the (as-yet-undiscovered) element 137 would be the end of the periodic table. Discoveries made since Feynman's suggestion about element 137 have shown that his idea probably isn't right, but element 137 will be an odd milestone on the periodic table for many reasons.

The number 137 is known in physics as alpha (α). Looking at it in a rather simple way, α is a dimensionless, fundamental physical constant with no units attached to it, and it defines the strength of the electromagnetic interaction in our universe. Numbers like the fine-structure constant are believed to be universal in nature; that is, they are the same throughout the universe and have and will be the same throughout time. The fine-structure constant itself can be most simply calculated by combining three constant values:

* The speed of light, c: 2.99792458 x 10^8 m/s
* The charge of a single electron, e: 1.602176634 x 10^{-19} Coulombs
* h (or actually, $\hbar$, which is Planck's constant divided by 2π): $1.0546c10^{-34}$ J•s

The variables are combined in the following formula:

$$\alpha = e^2/(\hbar)(c)$$

Substitute the values for the letters and you get a rounded value of 1/137, which, for simplicity's sake, is referred to as 137 (which is technically a^{-1}, but 137 is so much nicer to say than one-one-hundred-dred-and-thirty-seventh).

And it shows up over and over again.

One instance that can't be argued away is when 137 comes up as evidence that our universe is "fine-tuned" for life. If the fine-structure constant were larger, electromagnetism would be stronger, atoms would be smaller, and our universe wouldn't exist as we know it. And if the fine-structure constant was weaker, atoms would be larger, and most likely unable to hold together. In other words, in an alternate universe where these three values aren't what they are in ours, life as we know it, perhaps even matter as we understand it, would not exist. Even if we could somehow travel to different universes, if we accidentally went into one with a different fine-structure constant, things would not end well. Upon entering the universe, all the matter, fields, and forces we would interact with wouldn't be the same. Falling—or being torn—apart into your constituent atoms and neutrons would not be out of the question.

Viewing 137 as a constant that allows for life as we know it in our universe has historically opened the door to the question, "If our universe is fine-tuned for life, who fine-tuned it?" Obviously, the fine-structure constant is often grabbed by creationists and intelligent-design advocates as evidence of a "creator." It's also a favorite of and given special significance and powers by pseudo-science peddlers as well.

Scientifically, however, there's no evidence of 137 being any kind of "signature," be it from a deity or a higher-dimensional alien kid playing a computer game with us in it while it waits for its mom to finish cooking dinner. Given that the fine-structure constant is tied to the values of the speed of light, the charge of an electron, and Planck's constant at the energies currently present in our universe, its value has changed slightly over the history of the universe. Given our understanding of conditions just after the Big Bang, the fine-structure constant would have had a value of approximately 1/128—but for only a few microseconds (if that) after things got going.

Even though the word "constant" is part of the name, not all physicists are sold on the idea that 137 is forever and ever the same everywhere throughout the universe. Researchers have pointed to data that seems to suggest that 137 may not be as fundamental or as constant as is believed, but those conclusions are somewhat controversial and are still a matter of debate. For something that has the word "constant" in its name, the fine-structure constant is still a phenomenon of physics that is under active research.

Given all of that (and given that its value is apparently constant), 137 is something that any sufficiently technologically advanced alien culture in our universe would have discovered on its own. Send out a message or a beacon that broadcasts "137" around the galaxy and you've put up a sign that says, "We're kind of smart—come on over!"

In terms of Rick and Morty and the multiverse in which they operate, the use of 137 wasn't an accident.

CHAPTER 9

Dark Matter

★ ★ ✦ ★ ★

Dark matter isn't good enough for Rick Sanchez. The mystery stuff that's all around in our universe is apparently all around the C-137 universe as well. Side note: I'm going to call it "stuff," even though that's not the most scientific-sounding word available. For our purposes, "stuff" is a real thing; matter, and other descriptors, are a little fuzzy.

But Rick wanted more than just dark matter. He invented "concentrated dark matter," and with that became a target for the Zigerions, the universe's most ambitious but least successful con artists. According to Prince Nebulon and his crew in "M. Night Shaym-Aliens!" Rick's invention allows for accelerated space travel. Unfortunately for Nebulon and the Zigerions under his command, they fell for Rick's fake recipe for concentrated dark matter, and the resultant explosion took out their entire ship.

As a result, there are only two things we can be sure of regarding Rick's concentrated dark matter: 1) if it exists, Rick knows how to make it, and 2) the recipe is most definitely not cesium, plutonic quarks, and bottled water.

Dark matter is something that we share with the universe of *Rick and Morty*, but while Rick has apparently figured it out (as expected) in his universe, we've just started to scratch the surface of it here in our universe. If it even has a surface that we can scratch.

In case you're confused, fear not—it won't be the last time as far as dark matter is concerned.

LIFE, DARK MATTER, AND EVERYTHING

For a start, the universe—our universe—is not what it seems.

The universe we know—what we experience on a day-to-day basis, the matter and energy that make up our everyday life—is only 5 percent of everything there is. The remainder is unevenly split: 68 percent of it is dark energy, and the other 27 percent is dark matter. Let that sink in.

Ninety-five percent of, well, everything is stuff that we're kind of clueless about, scientifically speaking. Sure, we have ideas, but still—we just don't know. And on the matter side of things, there's between five and six times more dark matter in the universe than there is regular matter.

To make matters worse, we don't know what dark matter is, and we have never seen it. The best we can do is look around the universe and see its effects. While you may have been led astray by the theory that something is real because we can see its effects (I'm looking at you, parents, and your stories about Santa Claus . . .), dark matter doesn't play by rational rules (and doesn't get less real as you get older).

"We know dark matter exists in several ways," says Sophia Nasr, a PhD student in cosmology at UC Irvine and science adviser on the SYFY series *12 Monkeys*. "One is that we understand how gravity works, and we have the perfect laboratory for that right in our cosmic neighborhood: the solar system. The sun is by far the dominant mass in the solar system, so the gravitational force planets experience will be due mostly to that. We know that the force of gravity decreases as the inverse square of the distance of the planet from the sun, meaning the velocities decrease as the inverse square root of the distance. So we know how things should behave in a gravitational field.

"But when we look at galaxies and see how they orbit about the galactic center—which usually has a supermassive black hole—we see something weird: stars orbit with relatively constant velocities throughout the disk! This tells us there needs to be more mass than we can see providing a gravitational push for these stars. If there wasn't dark matter holding these galaxies together, these velocities would cause galaxies to fly apart!"

In short: there's just not enough visible mass to keep the galaxies doing what they're doing in the way that they're doing it. What's allowing them to keep on keeping on? Something that has a lot of mass (so it's a lot of matter) that we can't see (it's dark). It sounds simplistic on its surface, but no other candidate could stand up to the evidence that was seen. For decades, this mystery of galaxies that spun without falling apart puzzled astronomers, with the idea of dark matter as the possible solution slowly growing in acceptance.

Another piece of evidence for the existence of dark matter comes from gravitational lensing. While we've previously talked about gravitational lensing seen with stars and galaxies, the rule—that mass bends space-time—also applies to dark matter as well. As Nasr says, "In lensing, we see things like clusters of galaxies with weird warped galaxies that look like arcs; these are actually galaxies behind the cluster, and their light is being bent by the mass of the cluster. Of course, we also know how much bending should happen depending on the mass that's curving space-time, and we find that, even when we account for all the mass of the stars and gas, it's not enough to explain the amount of bending we see the clusters doing to space-time."

The third way we know dark matter is out there is actually the most convincing, and it happens when two galaxies collide. When galaxies collide, it's actually nothing really that special. The stars just keep moving. There's a lot of space between stars, so the whole "collision" thing is a little misleading when it comes to them. But galaxies aren't just stars, they're also gas, and with that, we can see evidence that supports the existence of dark matter.

Take the Bullet Cluster, for example—two colliding clusters of galaxies where the stars have all moved past each other but the gas is trailing behind. "It's important to understand that in galaxy clusters, out of the visible mass, gas actually accounts for far more mass than do the stars," Nasr says. "So, with the gas trailing behind the stars, there should obviously be more mass in the gaseous region than with the stars. But we find that's not true: the stars are far more massive, which means there's dark matter that moved right through the colliding galaxies with the stars!"

And each cluster of dark matter moved through the other as if it didn't exist. This giant, massive stuff that we can't see or measure is dark matter.

But if dark matter is everywhere and has effects that can change the shape and directions of whole galaxies, you would think it would be at least a little more noticeable. Well, that's actually right in line with our perspective on the larger galaxy and universe. When we look at our galaxy as a whole, our solar system isn't even noticeable, so it kind of follows logically that we wouldn't expect to see the effects of dark matter, even though the individual particles are probably chugging around you (they move slowly compared to other invisible things that can move through you, like neutrinos, which we'll talk about later) and even passing through you. There's just not enough here to cause any changes, and how could it? The sun dominates the mass of the solar system, so it calls the shots on a solar-system-sized scale.

But that's on a scale of you and the solar system. Our solar system has a diameter of about 1.58 light-years, measuring out to the Oort Cloud, while our galaxy, the Milky Way, has a diameter of 100,000 light-years. Dark matter's effects aren't that noticeable at the solar-system scale, but on the galactic scale, there's enough that the cumulative effects can be seen and felt in galactic spin, lensing, and the like.

So dark matter has mass and exists. Is that about all we can say for sure about it?

Not exactly.

WHAT IT IS

Dark matter is a thing, but the thing that it is, and its specific "thingness," makes it really, really tough for the rest of us things to do anything with it. It's easiest to think of a neutrino, which is a subatomic particle with very little mass and no charge. A neutrino is an electron's less interesting cousin, but dark matter has one significant difference when compared to neutrinos: it's "cold," while neutrinos are "hot." As Nasr puts it: "When we say dark matter is 'cold,' what we really mean is that it's nonrelativistic—that is to say, it's slow. Neutrinos, on the other hand, are superfast and relativistic, which is why they're 'hot.' So [dark matter is] everywhere, but it's not zipping around like neutrinos are."

One reason dark matter is so difficult to study is that it doesn't interact with anything except gravity, which explains why it shows up in galactic spin, lensing, and galactic collisions, but nowhere else. There are four fundamental forces in the universe—gravitational, electromagnetic, weak nuclear, and strong nuclear—and regular matter interacts with all of them. Dark matter, however, dances with only one. Like regular matter, dark matter has mass, so therefore it can interact with and feel gravity. That's it.

This actually helps explain why we can't see it: because it doesn't interact with electromagnetic forces, it doesn't give off or reflect any electromagnetic waves, which includes radio waves, infrared radiation, microwaves, ultraviolet radiation, X-rays, and gamma rays, as well as the entire spectrum of visible light. Without any of these, there's no telescope of any strength or magnitude that will be able to "see" dark matter. You can't spot dark matter directly, so you can't go looking for it directly—you need to look for interactions between dark matter and visible matter. But that's challenging, since it seems that regular, visible matter can't seem to touch dark matter, and dark matter doesn't seem to interact with itself too much. That's not to say interactions are impossible, though.

"The key to looking for signals of dark matter interaction is to understand what kind of interaction you're looking for," Nasr says. "For example, many dark matter detectors way underground are full of liquid xenon, and they sit there hoping to see dark matter interact with a nucleus, which they'd see by the electrons and photons coming out of the interaction. Other ways include hoping that the Large Hadron Collider in Geneva produces an interaction. But we need to be looking for the right signals; we might not be, and maybe that's why we haven't detected it yet.

"Worse yet, maybe dark matter doesn't interact with the particles we understand at all, and has interactions only with what we call the 'dark sector,' a sector that includes particles we can't see, including dark matter. If dark matter interacts with itself with a different force than those we understand, then that would be a 'dark force,' an additional particle in the dark sector. If this is the case, we aren't necessarily doomed; maybe one of the particles in the dark sector will interact with the particles we understand."

Researchers are hard at work, trying to identify exactly what dark matter might be. Within the field of candidates, there are a couple that are more widely known: WIMPs (Weakly Interactive Massive Particles), things that have mass but also very weak interactions with matter; and axions (hypothetical, light, neutrally charged subatomic particles introduced in 1977 as a means to solve other nagging problems in a branch of quantum physics called quantum chromodynamics).

"But there are other possibilities," Nasr says, "like if there's a sterile neutrino, or, if there are extra dimensions, then dark matter could be particles that got stuck in those dimensions while the universe cooled down, and now they appear massive [as in, having mass] because there's momentum in an extra dimension we can't access, and we also can't see it because we can't access those dimensions. Or it could be self-interacting dark matter (SIDM)—this is one type that would have a 'dark force' particle."

Or, dark matter might not be any of these. The hunt for dark matter takes place at the edge of what we know about physics.

"Discovering dark matter would mean discovering new physics—this is physics beyond the standard model," Nasr says. The standard model contains all the subatomic particles we think are responsible for matter and all its interactions. "Dark matter is not predicted in the SM, so it doesn't fit in anywhere. That's why it's so exciting! And while you might think that means dark matter may not actually exist, neutrinos are actually in and of themselves proof that the SM has problems: the SM predicts massless neutrinos, but they actually do have masses, albeit very small ones."

So the standard model may be due for an overhaul, and this new version may just include dark matter. It's a real part of our universe, and models of the fundamental particles should include it.

WHERE DOES RICK FIT IN?

If we keep with our base assumption that Rick knows just about everything in his own universe, then he would understand dark matter. Could he "concentrate" it? Even asking that is like going down Hypothetical Avenue and turning left onto East Hypothetical Boulevard . . . but probably no. Even if he did understand its properties.

Given that dark matter candidates are all subatomic particles, the idea of "making" it is a little far-fetched, and has an issue with the Law of Conservation of Matter, which strictly forbids you from making new matter—and subatomic matter, dark or not, is about as fundamental as you can get. Concentrating it? It's so, so diffuse. Even if you could collect it (it doesn't like to interact with our matter, which is what your collector would be made of), you'd have to scoop up a galaxy's worth for it to be of any real use.

But Rick's endeavors might not be completely fictional.

"If we could figure out a way to make [dark matter] interact with something, then we could, in principle, have an almost limitless amount of fuel in the universe for spaceships to be able to use, since they're always going through it," Nasr explains. "It's far more complicated than that, so a story like this would need to cover all these grounds. But yeah, that's how I could see it fit in, without just being there and holding galaxies and stuff together in space."

So the mystery continues. The universe we see is only 5 percent of the total of what is really there. Dark matter makes up about five times the matter that we know, see, and interact with, and if we're going to say one day that we understand the universe around us, we probably should keep working on that stuff that we can't see or interact with as well as the stuff we *can* see.

FINAL WORD: DARK ENERGY

While not mentioned in *Rick and Morty*, dark energy is the other big unknown in the universe. As mentioned earlier, matter makes up about 5 percent of the universe, dark matter makes up 27 percent of the universe, and dark energy makes up the remaining 68 percent. While it's easy to conflate dark matter with dark energy . . . don't. Dark matter is stuff; it has mass and interacts with gravity. Dark energy is just that—energy. It does things. It, as best we can tell, makes things happen.

Dark energy was discovered through observations that our universe is expanding, and expanding at an ever-increasing rate. Everything is moving away from everything else, faster now than it was when you started reading this book. Yes, gravity still works, and the fundamentals are still in place, but on the cosmological scale, something is pushing everything apart from everything else, and we have no idea what it is.

This is not an artifact of the Big Bang, and things are going to slow and snap back once they reach a certain point, like a rock thrown upward falling back to Earth once it reaches its highest point. Stars and galaxies have been observed to be moving faster as time goes on, not slower—as if that rock you tossed up into the sky just keeps going up and up and up. Just like the rock eventually disappearing from sight, someday in the far, far future, galaxies outside ours will have completely receded from view. We'll still be a part of, and see, the Milky Way, but our galaxy will be alone. All the other galaxies will have moved so far away from us that not even their light will reach us anymore.

While it's not exactly productive to discuss which "dark" branch of physics has had more luck in figuring things out (probably dark matter, but who's counting?), that's not to say that there aren't ideas about where dark energy comes from, with the quantum effects of the vacuum of empty space playing a major role.

Regardless of what it is, dark energy, like dark matter, will need new branches of physics to explain.

IS CONCENTRATED DARK MATTER EVEN A THING IN RICK'S UNIVERSE?

Concentrated dark matter could be a red herring.

Go with me here: if you have portal technology and the ability to instantly transport yourself from any point in your universe to any other point in your or any universe, why do you need accelerated travel? Especially accelerated travel that uses dark matter, which is pretty complicated to understand, find, grab, and use?

For a show that loves to maintain its internal logic, having concentrated dark matter as a sought-after substance is a little odd. If portals will get you or your ship where, or near where, you need to go, why do you need, basically, a really fast car?

There are two possibilities that come to mind: one, the Zigerions don't have portal technology, and need to move faster in space for trade and other reasons—but given the size of space, even having really fast travel barely makes you an intergalactic species, and that's only if your neighbors are also advanced and living nearby. Or two, the Zigerions have managed to colonize a corner of the galaxy but don't yet have portal tech, aren't aware of portal tech, and aren't the sharpest alien species in the drawer. Perhaps they've seen Rick in a few places when he was portal-ing around in their space, and the time frame in which they encountered him would suggest that he was moving extremely fast (again, they don't know what portals are).

The other possibility? Rick made it up. Maybe, in a conversation with the Zigerions, Rick said something offhand about being able to move extremely fast thanks to "concentrated dark matter," something that doesn't exist, but hey, he wasn't about to give these loser aliens access to portal technology. And ever since, the Zigerions have believed Rick's lie about concentrated dark matter. That doesn't say great things about the Zigerions as a species, but they are, after all, the least successful con artists in the galaxy.

CHAPTER 10

Memory, Dreams, and the Brain Playground

* * ⭐ * *

If you were asked to think about your strongest memory from your school years, what would it be? Not the specifics of who and why and what you really thought about that other student, but the quality of the memory itself. Is it a narrative story, replaying like an episode of a television drama in your head? Or is it something different, like a collection of sensations—colors, faces, images, sounds, smells—that your brain stitches together to make a coherent memory?

The longer you think about memory, the weirder it gets. Some memories are vivid and can elicit an emotional reaction, but others are hazy or incomplete, or simply aren't there at all. They're a patchwork of old, new, crystal clear, muddy, absolutely true, and not very true at all.

These questions keep researchers (and maybe now you) up at night, or at least motivated enough to keep probing the huge mystery that is memory.

Rick's already done it, though. He's able to implant false memories and ideas in people's dreams, construct fake memories in his own mind, and even remove traumatic memories from his grandson's brain. But looking at Rick's understanding and application of the science of memory, it's apparent that this is one of those places where our science and understanding of memories is pretty similar to Rick's, even if we don't have the same applications (not yet, anyway).

Memory is a tricky subject, and the more they study it, the more researchers find that much is persistently unknown or stubbornly refuses to reveal itself. But the science is advancing quickly, and some of what we now know about memory would probably slow Rick down, if only for a minute.

WHAT IS MEMORY?

As we take in new information through our senses, our brain does its level best to organize it. Initially, memories are stored as their immediate, raw sensations in a sensory stage. From there, the experience is encoded and converted into a construct. More intense experiences cause neurons to fire with a higher frequency, increasing the chance that the event is encoded as a memory, while less intense events are moved into short-term memory. According to recent research, that storage is extremely limited and can manage only about six or seven items for about thirty seconds, tops. Repeat something over and over, like the telephone number of that hot guy or girl you met at the store, and you're constantly starting the clock over again for your short-term memory.

It's the best of the best, or the most intense, repeated, and utilized stimuli that get from short-term to long-term memory, and as far as we know, barring accident or disease, long-term memory can hold an unlimited amount of information for a lifetime. And those memories are actually physically hardwired in.

In the late 1960s, a better understanding of the physical basis for memory led Dr. Eric Kandel at the New York University Medical School to look at an extremely simple nervous system that was capable of producing memory—that of the sea slug *Aplysia californica*. The slugs' nerve cells were very large and easily dissected, and responded well to conditioning with stimuli. Kandel and his associates constructed a simple nerve circuit made up of a motor neuron and a sensory neuron, and found that when the sensory neuron was stimulated it released messenger RNA (mRNA), which caused it to grow new connections, called synapses, with the other cell. This was the first proof that memory involved a structural change in the nerve cells of the brain, and ultimately won Kandel and his co-researchers the Nobel Prize in Physiology and Medicine in 2000.

Following what Kandel found in sea slugs, the formation of a memory goes like this: when separate sensation inputs (touch, smell, visual, auditory) are received, they're routed to a part of the brain called the hippocampus, a horseshoe-shaped structure located in the temporal lobe. The hippocampus screens the inputs and decides whether they're worth remembering, and if they are, it consolidates the separate inputs into one single memory, classified by neurologists as a long-term memory.

Pieces of the memory are "stored" in different parts of the brain: the visual elements in the visual cortex, the sounds in the auditory cortex, smells in the olfactory cortex, movement in the motor cortex, and emotional sensations in a structure called the amygdala, as well as structures associated with certain stimuli. In this way, an actual, physical memory isn't so much a "spot" in the brain as it is a web of connected neurons weaving throughout the brain called a "memory trace." Sometimes these webs are associated with others that have stronger synapses, which are the connections between individual nerve cells. At these connections, a stimulated nerve cell releases a chemical called a neurotransmitter, which lands on a receptor of the nearby nerve cell and thus causes that nerve cell to become stimulated and to pass the stimulation via neurotransmitter release on down the line. Some synapses are "stronger" than others due to an increased number of receptors for the neurotransmitter on the receiving neuron.

For example, if you have to sing the entire alphabet song in order to remember that P comes before Q, you're accessing an associated web in order to get to the memory you want, but cannot stimulate by itself. The memory of the song and the ability to sing it in your head had stronger synapses than the simple fact that Q comes after P.

This initial storage of memories is called "synaptic consolidation," while "system consolidation"—turning a fresh memory into a long-term one—occurs over a period of weeks, months, or years, as the neural pathway, through repeated use, is established as an independent memory, no longer under the control or influence of the hippocampus.

These new connections can be reinforced by repeated use, like studying a topic or practicing an activity; and they can be minimized by neglect, which is how we forget things we previously knew. Long-thought but forgotten theories can occasionally spring back to mind due to a random stimulation that provides enough activity for the memory's trigger neuron to fire, while some memories are most likely gone forever, as underused connections are ultimately let go by the neurons and absorbed by the surrounding brain tissue.

But remember: as a new memory is formed, neurons are forming new connections with one another. This is due to a feature of the nervous system called neuroplasticity. This can be seen in many instances, like when a stroke patient relearns how to walk, despite a portion of the motor cortex being damaged. The brain constantly rewires itself to meet the demands of the user.

While the brain is the most important overall organ in the body (according to itself), it's the hippocampus that is crucial to the formation of memories. Think of it as the gateway to memory. Damage to the hippocampus, either physical—such as with a lesion from trauma—or physiological based on brain chemistry—as happens in Alzheimer's and other cognitive diseases—can cause amnesia, ultimately denying the victim access to old memories or the ability to create new ones.

So, in the brain, memories have an actual, physical location in the increased connections between neurons, set up in specific patterns that connect multiple parts of the brain. That's laying memories down. As you can probably guess, accessing those memories, or just remembering itself, involves these same new pathways.

If you're getting uneasy because you know you sometimes forget things, don't start worrying that you have brain damage just yet. "Forgetting" information that you know you once knew can actually be caused by a variety of things: the input may not have made it out of the sensory phase, it may have died in short-term memory, or it was otherwise not deemed "important enough" by the hippocampus. Or you just might not be able to retrieve the accurate information.

In light of this last point, let's look at how we remember. To recall a memory or a piece of information, the same network of neurons activates. The act of remembering, of lighting up that collection of synapses, reinforces the connections between the individual neurons, requiring mRNA and subsequent protein synthesis and building of new connections—either reinforcing or tweaking a small part of the memory. Research has shown that during recall, memories are vulnerable to modification due to the "rebuilding" aspect of remembering. In the lab, memories can actually be wiped out by giving the test animal anisomycin, a drug that blocks the formation of the proteins that are needed to build and reinforce new connections. Put an animal in a situation where it has to remember something it's done before, dose it with the drug, and it literally cannot remember what to do. The memory is gone, and it doesn't come back.

Neuroplasticity giveth, and neuroplasticity taketh away.

MEMORY AND DREAMS

Given their similar nature, we've long placed roughly equal importance on dreams and memory, and research seems to underscore their similarity, to the point that one can't happen without the other. According to research, sleep seems to help with the consolidation process of memory acquisition, "firming up" the memory within the structure of the brain. There's been a lot of research that shows test subjects—from mice running mazes to college students studying for finals (and looking to make some extra bucks from a psych department study)—retain more information when they get a decent night's sleep. This close sleep-memory relationship is also thought to be the reason why many, if not all, dreams include strong elements from memory, from as recently as the previous day to as deep as from childhood.

During the night, people move through two phases of sleep, one characterized by rapid eye movement (REM), which is a lighter sleep and the stage when dreams are most likely to occur; and deeper, non-REM sleep, also called slow-wave sleep (SWS). Studies have shown that memories go through a type of "processing" during SWS, with essential memories being marked for retention and low-information memories being let go. During REM sleep, essential memories are changed so they can be held on to as long-term memories. This connection between sleep and memory is the focus of current research sponsored by DARPA (Defense Advanced Research Projects Agency, part of the US military complex), which has shown that individuals who receive low-level electrical stimulation during specific phases of SWS showed better integration of information to which they were exposed just prior to sleep, which begets better retention.

Other research suggests that memories are disconnected from any intense emotions with which they're associated during sleep. Once the emotions and the memory's images are split, the memory is filed for consolidation, while the emotion evaporates. That's not to say that memories can't cause a resurfacing of emotions; just that the memories themselves are stripped of emotions, and any emotion you feel with a memory is one that you spun up in relation to the memory, rather than its having been stored as part of the memory.

At least one line of thinking suggests that emotions that remain attached to memories are quite literally "nightmare fuel," as the sleeper is once again experiencing an intense emotion, but their conscious mind is unable to process it, leaving it to the unconscious mind, which is basically still a scared primate on the plains of Africa, to react. This reaction bubbles up through our unconscious, picking up some nice attachments along the way, and suddenly the anxiety you felt at a meeting earlier in the day becomes a clown chasing you through an abandoned amusement park at night while your shoe is untied. You're also naked, which is much worse.

While memories seem to be coupled to sleep and dreams, dreams themselves are a chunk of the human experience that is still not fully understood, though that hasn't stopped us from trying to understand them. Freud felt that dreams, as the pathways to our unconscious mind, were loaded with symbolism. Dream interpretation can be yours for a couple of clicks on the Internet. And as a culture, our stories are full of characters for whom dreams reveal secrets, or who are guided, for good or ill, by their dreams.

But back on the science side, dreams may play a role in memory processing, and it may be true that you have a tough time remembering dreams because all the stimuli are already inside your head, so there's nothing new for the various brain cortices to work with (no real smells, sounds, touches, tastes, etc.), but that's not all we know about them.

Common lines of thought about why we dream include that dreams may be the means by which the brain "defrags," or unwinds after the conscious mind has its run. And while the consciousness is away, the unconscious mind will play. Dreams themselves come in every shape and size, and if you ask a hundred people from around the world about the type of dream they had last night, well . . . okay, probably a lot would be about sex, but there would be a broad collection of topics, themes, settings, characters, story lines, protagonists, and antagonists. Studies suggest that there will be several people who have uncannily similar dreams; as a species, we seem to be haunted by the same brain-ghosts at night. And also, virtually all the dreams would follow a narrative, at least in the most basic cause-and-effect sense. Things happen in an order, with the earlier thing most often causing the later thing.

Unlike in deep sleep, much of the brain is active during a dream, and the centers of dream activity seem to be in a portion of the prefrontal cortex and medial temporal lobe system. The motor cortex, which allows for movement, is a notable holdout—which is a good thing, because otherwise you could end up acting out your dreams. Some parts of the frontal lobes are quieter as well, which is thought to limit logical, critical thinking during dreams, as well as your internal censor, which, when you're conscious, keeps you from saying or thinking things you probably shouldn't. This all means that you don't find anything wrong with your deceased grandfather talking to you on a beach while you're both wearing inflatable *T. rex* costumes and watching a beach full of people and animals have sex. We all can be pretty "live and let live" in our dreams.

If we follow the anatomical path of a dream, it seems that the lower brain stem, the more "primitive" part of the brain, activates first, and much of the upper part, the cortex, then wakes up in response and does its thinking in an organized way. Somewhere between those two, dreams happen. It may be that the upper brain is trying to make sense of the sensations from the lower part and just brings some memories and associated emotions along for the ride, or perhaps the two parts act in concert to build a dream. It's still a mystery.

One thing that's not a mystery is the importance of dreams. Numerous studies have shown that without REM sleep and dreams, memories are not consolidated and are lost. Several links have been made between lack of REM sleep (and therefore dreams) and mental health issues such as depression, schizophrenia, and personality disorders. More and more clinicians point out that sleep deprivation, which much of the modern world seems to view as a badge of honor, is also dream deprivation. And that's not good.

Okay: dream groundwork laid. Let's *Inception* this thing.

"Lawnmower Dog" took playful pokes at the film *Inception*, which Rick hates (though he then went on to do the exact same thing that happens in the movie). In reality, influencing someone's dream is possible, just not in the sexy, *Inception*-esque, earplug way. Planting suggestions for dreams in our world involves presenting topics to dream about (either directly or indirectly) or supplying stimuli to the individual while they're asleep, such as low auditory input or faint visual or tactile input, all at a threshold low enough that the dreamer will not wake.

While putting ourselves into the dreams of others is a science fiction fantasy for now, there are some interesting developments that could be the first steps on the road to an *Inception*-style reality. When an individual's consciousness is active, their neurons fire in specific patterns. They do some mathematics, and a set of neurons fires. They think of their mother, and another set fires. If you place the person in an MRI scanner and record their brain activity as they think of a variety of subjects, and then you let them sleep in the MRI scanner, you can watch for the same patterns as before to see what they're dreaming about. It's a crude method with very low "resolution"—you wouldn't be able to play their dreams on a movie screen or anything—but you would have some insight as to what the person is dreaming about, and from there the dreams could be manipulated in the manner mentioned above.

Getting into someone else's dream and observing or playing a role in the action isn't really possible with today's technology, or with our understanding of dreams and the brain. There was, however, something in the homage to *Inception* in "Lawnmower Dog" that had a little research and science behind it: the seeming dilation of time in dreams. When investigating the passage of time in dreams, Dr. Daniel Erlacher at the University of Bern in Switzerland asked individuals who are able to stay "awake" in dreams (so-called lucid dreamers) to perform specific tasks in their dreams that were time-dependent.

Through an experiment that allowed Erlacher to confirm that his sleepers were doing as he had requested rather than faking being asleep, he found that in "dream-time," activities took a longer amount of time than they did in the waking world—sometimes up to 50 percent longer. This was a reverse of the complicated math of the multiple dream worlds with their own time frames in "Lawnmower Dog," where time moved faster the deeper down Rick and Morty traveled, and it does seem to hint at something many of us have experienced: that time can happen at a slower pace in dreams. Whether this is due to delayed processing by the brain or something altogether different still needs research.

But there's one lingering question in all of this: If our dreams are fed by a blend of our memories and thoughts, then why was Summer in Mrs. Pancakes's dream? Why would a fictional character in Mr. Goldenfold's dream be dreaming about Summer as a guest at her pleasure palace . . . unless that wasn't Mrs. Pancakes's dream and was actually part of Morty's, and . . . oh, never mind.

CHANGE 'EM

Memories are, at their most basic, an alteration in the structure of the brain. Altering them isn't a big deal.

As mentioned earlier, to form a new memory, a new connection must be grown between neurons in the brain. When the memory is recalled, a signal is sent to reinforce the memory, meaning that the proteins that helped to build the connections in the first place are again released. It's in that time frame that the memory is vulnerable and relatively easy to change or modify.

The approach to altering memories in this fashion is called "reconsolidation" and refers back to the formation of the memory in the first place, as well as the above-mentioned vulnerability—a period of time when the memory itself is unstable and can be modified. The new information helps to lower the extreme emotional response to the memory, and as a result, the next time the memory is recalled, the emotional response will be lessened.

Reconsolidation can and has been used to treat people suffering from conditions such as arachnophobia and severe posttraumatic stress disorder (PTSD). In therapy, the patient is asked to recall the situation that causes the emotional response and the memory that's attached to the emotion. Now that it's been recalled, the memory is pliable. The emotional response that the patient has upon recall can be treated pharmacologically at that point (while the memory is fresh) with drugs that reduce the adrenaline response, or the emotional response can be approached using methods more psychological in nature, such as the visualization of being an observer to the event rather than a participant.

The goal of this treatment is to reduce the emotional response that's attached to the memory so that the next time the memory is recalled, it doesn't bring up as much emotion. In best cases, the treatment can permanently modify the emotional response while leaving the original memory intact. A 2017 study published in the journal *Psychotherapy Resources* reported that when sixty-five volunteers suffering from PTSD were treated with reconsolidation therapy, 71 percent lost the official diagnosis, and of those, 65 percent were in complete remission.

So, memories can be tweaked and altered, and the emotions associated with them can be dialed down. In case that's not good enough, though, don't worry—you can just make up memories altogether.

FAKE 'EM & MAKE 'EM

While sitting in the Denny's of his mind in "The Rickshank Rickdemption," Rick walked Cornvelious Daniel through a traumatic memory of how he invented portal-gun technology. The Federation agent followed along as Rick sold him on the story—including an added detail about McDonald's Szechuan sauce—which involved love, hope, and, ultimately, loss. And it (maybe) was all fake—a completely fake memory designed to be 100 percent believable. That's so Rick.

Rick's artificial memory was a beautiful example of something that can be done with virtually anyone. You can even do it to yourself.

It goes like this: in a research setting, a subject is asked to remember a time in their past. The questioner has some information about that time prior to the session and can direct the flow of the conversation. As mentioned previously, memories are vulnerable when they're being actively remembered; therefore, a particular suggestion can be implanted in the subject, so long as it's done carefully and the new memory is seated in believability.

All the questioner has to do to achieve this is gently blend in a couple of real elements from the memories of the subject with the false memory and then ask the subject to remember them.

To further help the new memory along, or to help if the subject is being resistant to the new memory, the questioner can ask the subject to imagine what could have caused them to do the action in the new memory, hint that other people have memories of them doing the action, or suggest that most people have some level of repression in their memories, so it's okay if it takes a little while for them to remember this new (and completely fabricated) element.

In one of the early experiments where this method was used, the research conducted by Dr. Julia Shaw of London South Bank University was halted because it worked too well. In her study, Shaw convinced college students to "remember" a crime they had been involved with years before—a completely fabricated story. With just a little coaxing like that described above, the students turned Shaw's idea of being involved in a crime from something that could not have happened to something that might have happened to something that did happen.

Add new information to the old memory in this vulnerable state and apparently the new, albeit false, information gets blended right in. And this can be done quickly and easily with individuals or with groups who have a common reference point in their memories. This fragility of memories is often cited as one of the means by which conspiracy theories about public events can spread; you're asked to remember an event, and as you do, you're questioned by the believer in the conspiracy theory whether you remembered x, y, and z happening as well. The seed is planted, and your brain does the rest.

Back to Rick and Cornvelious Daniel: while walking down Rick's memory lane of portal gun creation, Daniel had no idea he was moving through something Rick had constructed, probably for this very eventuality, to protect the secrets of portal technology. All Rick would have had to do is come up with the new memory, blend it with some actual memories (was the placid and peaceful

Rick of his memory actually C-137?), and repeat it over and over and over until the real memory was gone, replaced by the new.

All of this helps explain why police and other authorities put so little weight on eyewitness testimony, and why there are many cases in which individuals convicted of crimes based on accounts from eyewitnesses are later conclusively proven by DNA evidence to be innocent. Our minds are powerful persuaders, and with just a hint of "You remember this happening, right?" and some reinforcement, the original memory can be altered to the point that it can never be recovered. Like it never happened.

SAVE 'EM AND REPLAY 'EM

We talk about memories being triggered all the time, from the traumatic to the silly. But current research suggests a way to record memories and play them back whenever they're needed—a little like what was seen in "Morty's Mind Blowers," but without removing the memories from the individual's brain.

At Columbia University, Dr. Christine Denny uses a method called optogenetics wherein she can "record" memories from genetically modified mice. This is done by inserting a gene taken originally from algae that codes a protein that responds to light into the mouse's DNA. When the mouse is treated with a specific drug prior to an experience, the gene is switched on, and any neurons that fire to construct the new memory embed the light-sensitive protein on their surfaces and become light-sensitive themselves. The particular memory trace is now made up of cells that have the light-sensitive protein embedded and will fire when light is shined on them.

To expose the cells to light and "turn on" the memory, Denny installs a surgical port in the mouse's brain capable of holding a fiber-optic cable with a low-power laser in its tip. When the laser is turned on, the light travels through the brain, activating the neurons with the light-sensitive protein, and the mouse "relives" the memory. Experiments with this method have shown that the mice, when placed in a new, brightly lit environment (that would naturally elicit fear), can be made docile and even curious when a recorded memory from a session in a comfortable, natural environment (a "good" memory) is turned on. In short, the mice act as if they're in a comfortable environment, even though nothing about the fear-inducing environment has changed.

Of course, since they're mice, we have no idea what the "good memory" actually entails—sights, smells, sounds, or just an overall pleasant feeling—but the mice are clearly reacting to some memory that's retrieved when the neurons are activated. Also, the "storage medium" of the memories for the mice is still the brain, and due to the invasive nature of the technique, a relatively limited number of memories can be turned on again at a given time. Now it's just a matter of technology catching up.

Given the findings of reconsolidation, it's unsurprising that other research has shown that this approach to memory recording and playback can lead to a "memory synthesis," where confusion of new memories with previous memories leads to the creation of false memories. This can mean that the subject treats a fear-inducing environment as a comfortable one, or treats a novel object as familiar. While the finding is interesting in mice, what's particularly fascinating is that false memories of this sort are also seen in humans with memory disorders, such as Alzheimer's.

While switching memories on and off is still limited to mice, some of the researchers in Denny's lab are trying to use the lessons learned from initial optogenetics work to treat diseases that affect memory, like Alzheimer's. Early findings suggest that, at least in genetically modified mice that develop Alzheimer's, memory traces for memories that were thought to be lost can be reactivated, and the mice demonstrate a recall of those particular memories.

The research seems to work well in mice, but any kind of poking and prodding in a human brain to track down particular memories to be "reawakened" is still decades away.

SHARE 'EM

To get into the idea of sharing memories with other individuals, let's bring back our slimy memory pal, *Aplysia californica*, which was so useful in determining the original finding that, at its core, memory is structure.

As a stimulus is repeatedly encountered by the slug, mRNA is coded in neurons and new connections between neurons are made. Go into that neuron, take that mRNA, and put it into another slug; now that second animal has the memory of what the first animal knew.

Moreover, repeat it with just neurons in a dish, and they show the same response when stimulated as they do in the organism itself. In short, the memory has been transferred. That research, according to lead researcher David Glanzman, suggests that this means "memory" isn't so much about the connections between neurons as it is about the changes the mRNA causes in the cell and its DNA. Glanzman's view is a competing one, which suggests memory is all about structure, but even he admits it's not absolutely clear what's going on as memory moves from one slug into another.

The science for both saving memories and sharing them is something that *Rick and Morty* plays with a lot, constantly swapping memories and even full consciousnesses from body to body, both between clones and between humans and insectoid aliens. This, of course, opens the question of implanting memories into the brain of a different species, and, more specifically, two species that may not have the same overall anatomy. Which brings us to the idea of the show's memory parasites.

The parasites create entirely new memories in the minds of the infected individuals, which would require modification of brain architecture to craft the new memories. Following along with what we've been looking at with memories, the parasite would require a viral-like means of infection to infiltrate a new host's brain. From there, the parasite could play with the connections and memories of the person, crafting an artificial memory that's a blend of the new and the familiar—Hamurai at a family cookout, for example. By the parasite's asking the infected host to remember their good times, the memory is recalled and strengthened, increasing its chances of becoming permanent and therefore locking in the infection. By only crafting memories of "good times," the parasites are probably feeding on or gaining energy from dopamine, a neurochemical that reinforces behaviors that bring pleasure, one of the pieces of the brain's built-in reward system. That's the system that can make you smile when you remember a particularly good time with a favorite person.

It's also interesting to note that the parasites can infect more than one person at a time, since everyone "remembered" Cousin Nicky and his catchphrase. While the infection made for a hilarious episode, it also played with interesting brain science.

This science of memory is also crucial to the dreams of believers in cryogenics, who hope that one day their bodies, frozen after death, will be thawed and they'll start life again, memories intact. Or that one day we'll be able to upload our consciousnesses into a computer. Memory research strongly suggests that both of those technologies are far, far away.

THROW 'EM OUT

We should all probably just admit that Morty has possibly the worst case of PTSD ever shown in an animated series, or at least in an animated series about a science-loving grandfather and his grandson.

As we saw in "Morty's Mind Blowers," some experiences are just too much for Morty to deal with, and many (hundreds? thousands?) times after experiencing such events, Rick removes the memory from Morty's brain, or Morty asks him to take out the memory of the traumatic experience. In doing so, Rick removes the memory from Morty's brain, and, for some insane reason, he keeps all the trauma, the upset, and the anger in a chamber under the garage.

While optogenetics and the research of Dr. Denny and others suggests that "recording" memories may someday be possible by mapping the precise pattern of neurons that fire when that memory is being accessed, the technology for removing memories is kinda already here, if just in mice and snails.

Methods of memory modification that can make the original memory obsolete have already been mentioned, but if losing a specific memory is the goal, there are a couple of ways to go, both of which depend on neuroplasticity. The more psychological approach involves something like reconsolidation—recalling and replacing the memory while it is open to change—creating a fake memory to paste over the original, as it were.

The more practical method is rooted in the physical and anatomical basis for memories. Deleting memories should be just a matter of editing the connections between the neurons that make up the memories. For that, researchers went back to *Aplysia* and began working with the slug's neurons, and were able to block the specific molecules that maintained a memory by decreasing the strength of the synapse (cutting the number of receptors for the neurotransmitter). More than that, they were able to selectively block memories associated with specific events (for example, seeing a red car when a crime is committed late at night on a dark street, causing the person to fear red cars), or nonassociated memories (for example, being out late at night on a dark street may not be safe).

The science and methodology are in the early stages and not even being discussed for human application yet. But the research continues, and maybe one day we'll be able to go full *Inception*. Or, rather, full Rick.

CHAPTER 11

Changing Smarts

★ ★ ✦ ★ ★

Your brain could—and should—be doing better.

At least that's an idea shared by Elon Musk, Mark Zuckerberg, Bryan Johnson, and Rick Sanchez.

After all, Rick turned the Smith family dog into a would-be world conqueror in "Lawnmower Dog." He also knows all about the cognitive side effects of Mega Seeds, and has definitely performed some kind of neural enhancement on himself. Brain functionality is right in Rick's wheelhouse.

And it's also in ours. Enhancing intelligence through science and technology used to be relegated to books like *Flowers for Algernon* or movies like *Planet of the Apes* or *Deep Blue Sea*, but now, tweaking your brain is a real thing.

BOOSTING YOUR SMARTS 1: NOOTROPICS

Remember how Morty shoved the Mega Seeds waaaay up his butt in order to sneak them through intergalactic customs? And remember, at the end of the episode, how he got temporary superintelligence and was able to recite the square root of pi and more, surprising everyone (except Rick)?

As Rick explained it, the temporary smarts were a side effect of the Mega Seeds being in contact with some of Morty's more delicate tissue. In other words, a chemical from the Mega Seeds made

Morty smarter, if only for a short period. You might think this is one of the farther-fetched elements of science fiction served up by *Rick and Morty*, but this technology? We actually have it right now.

"Nootropics" is the name given to drugs, dietary supplements, and other substances that claim to improve cognitive function—specifically, brain functions such as creativity, motivation, and memory, collectively known as "executive functions."

While some nootropics have shady origins, efficacy, and acquisition, others are prescribed by doctors, and a few are available on the shelf of your local grocery store. For example, millions upon millions of people dose themselves with a cognition-enhancing drug every single day: caffeine.

Stimulants of the central nervous system, like caffeine, are by far the most popular cognition-enhancing drugs, but are just one type of drug that can be classified as a nootropic. The research around them supports the idea that, in small doses, they do have a nootropic effect. This is achieved by the stimulant's ability to stand in as a neurotransmitter and activate either D 1 dopamine receptors or alpha-2 adrenoceptors in the prefrontal cortex of the brain, the part of the brain responsible for higher-order thinking. By stimulating the dopamine receptors, pathways involved in working memory and motivation can be activated, while the adrenoceptor activation can also help working memory as well as increase functions affecting attention and attention span.

The list of stimulants that show nootropic effects includes drugs as "mild" and socially acceptable as caffeine and nicotine, but also drugs aimed at ADHD and sleep-disorder treatment, such as eugeroics like armodafinil and modafinil (often referred to as the "Limitless pill," after the film of the same name), methylphenidate, and various kinds of amphetamines, including Adderall. Clinical medicine's effectiveness as a nootropic is limited, mostly due to the fact that it's tough to convince doctors that "poor cognition" is an illness that needs to be treated medically. Unnecessarily messing with brain chemistry isn't seen as a desirable avenue of treatment.

Mostly, the "results" of nootropic experiments come from "biohackers," individuals who seek to make their brains and bodies function better and use them as laboratories. Many obsessively document their experiences and make their results public. While many testimonials for various nootropics often read as if they're too good to be true, it's important to note that none of the nootropic drugs are, in any way, adding information to your brain or your learned skills.

For instance, if you don't know how to calculate the square root of a number, you're not going to suddenly be able to calculate it after taking a nootropic. The stimulants may help with laying down memories of information consumed while under the influence of the drugs, but they cannot magically gift the individual with sudden knowledge or abilities. They just help you work better with what you've got, through a combination of boosting awareness, focus, motivation, memory retention, and recall and stabilizing your mood.

While a handful of drugs and supplements claim to have nootropic effects, this is a field full of unregulated ingredients and unproven tests with results that cannot be clinically demonstrated. Ginkgo biloba, lion's mane, L-theanine, and other supplements are advertised as having positive cognitive effects, but there's precious little—if any—data to support those claims. In the world of nootropics, it's often difficult to tell the legimate researchers from the hucksters.

However, in case you were thinking about trying this at home, many nootropics can have pretty lousy side effects, as when Morty began seizing and drooling after the effects of the Mega Seeds wore off. There are no free lunches when it comes to altering brain chemistry. Case in point: stimulants. Low doses have been clinically demonstrated to have cognition-boosting effects, but high doses can temporarily reduce cognitive function or, in the case of the more powerful stimulants, lead to addiction, injury, or death.

It's always worth remembering how *Flowers for Algernon* ended.

BOOSTING YOUR SMARTS 2:
HUMANS + MACHINES

Rick was stopped from boarding the ship traveling through the wormhole in "The Whirly Dirly Conspiracy" because he had multiple Class C or above cybernetic enhancements. Check the screen of the security monitor and you can see that there's something highlighted in his skull.

Brain-computer interfaces (BCIs) will be explored in-depth in the next chapter as we look at prosthetics and how BCIs can be used to allow the brains of individuals with specific impairments to communicate with those devices, such as robotic arms, legs, and computer interfaces for communication. While there is crossover between the two fields, in this section we're going to look at mixing a human brain with a computer purely for reasons of user enhancement—a full-on blend of the biological and the technological.

Human brains are fragile and sometimes temperamental things, prone to faulty wiring, disease, and issues with recall, retention, reasoning, and judgment. Why not get help, or at least offload part of the work at a crucial point—whether when we're a contestant on *Jeopardy!* or taking a make-or-break final in a critical university course?

The idea of humans with computer brains (either entire or partial) has been a staple in science fiction for decades, but now reality is catching up to the fiction. Along with the BCIs crucial for advanced prostheses and other applications, neural-computer interfaces are a very active area of research and experimentation, with players from a variety of fields.

The mixing of the biological and the technological starts on common ground: electricity. First, a quick refresher on how your nervous system works: neurons in the brain communicate via very small electrical signals generated by imbalances between potassium (K+), sodium (Na+), and chloride (Cl-) ions. K+ ions are selectively pumped into the nerve cell, while Na+ ions are pumped out in a ratio of two K+ in to three Na+ out, resulting in more sodium and chloride ions outside the cell and more potassium inside. The difference in the concentrations of ions results in an electrical difference between the inside and the outside, which is called the "resting membrane potential." In most nerve cells, the difference between potassium and sodium ions keeps the resting membrane potential at about -70 millivolts (mV). In other words, the charge inside the neuron is about 70 mV less than the charge outside.

If a stimulus signals the nerve, the resting potential starts to change, and sodium channels in the cell's membrane open up, letting positively charged sodium ions flood in. This causes a "depolarizing" current, where the previous -70 mV value inside the nerve cell starts to become more positive and head toward 0 mV. When the charge (called the potential) inside the cell reaches -55 mV (called the threshold), the cell fires a signal called an action potential. Ultimately, the charge inside the cell will change from -70 mV to +30 mV. At this signal, sacs called vesicles (which hold neurotransmitters) release their goods, allowing the chemicals to move across the gap between neurons (called a synapse) to stimulate the next neuron down the line via the same process. After firing, the nerve cell resets and awaits the next stimulus that will form the next action potential.

Anything you think, feel, taste, touch, see, or hear is all just a string of electrical signals moving from neuron to neuron.

The electrical nature of our nervous system can easily be seen in the use of transcutaneous electrical nerve stimulation, commonly known as TENS therapy. In this approach, low-voltage electrical signals are sent through the skin to cause muscle contraction or a blockage of nerve signal transmission from a site of local pain to the brain, resulting in a decrease or cessation of the pain. The electrical nature of the nervous system is also why Tasers work as well as they do: they effectively scramble the signals being sent from the brain and spinal cord to the skeletal muscles, and movement becomes impossible.

Initial work on functioning neural implants was performed in 2011 by Theodore Berger of the University of Southern California and a collaborative group headed by Sam Deadwyler at Wake Forest University. The scientists were able to place small chips in the brains of rats and record information from their hippocampi as the animals learned a new task and stored the information. Then, using a drug, they shut down a part of the rats' brains, while another device, which could "play back" the signals from the earlier "recording," was wired into their brains. The implant system allowed the rats' brains to encode the long-term memories of the event and increase the length of time the memory was retained. Berger and his colleagues saw improvement in recall functions in both rat and monkey models.

Further progress with brain implants is being made in relation to treating diseases and brain disorders. Many approaches are looking at ways to strengthen memory in individuals with Alzheimer's or other neuro-degenerative diseases. In another small study at Wake Forest, researchers were able to record simple memories from human test subjects as they were being made, and then were able to replay the neuronal signals at a later point, boosting the subjects' short-term memory.

Other versions of computer "helpers" for the human brain are being worked on by several groups, including BrainGate, a consortium of researchers from Brown University, Case Western Reserve University, Massachusetts General Hospital, Stanford University, and Providence VA Medical Center. The BrainGate implants, which can be placed on the surface of the brain, look like tiny, thin hairbrushes, with the bristles being microscopic electrodes that penetrate specific regions of the brain. The implant can stimulate and be stimulated by certain regions of the brain.

The idea of an implant loaded with a huge amount of data, such as the sum contents of Wikipedia, is still a little way off, however. Intelligence, as well as memory, requires many different parts of the brain to act in concert. With the implant, information may be "in" the brain and accessible, but the interpretation and the application of the information in context with other knowledge would be quite different from just recalling raw facts. True understanding of information comes from careful consideration, reflection, and fitting it in with previously held knowledge.

While the information from an implant may make one person think of one thing when exposed to the information it contains, another person may make a wholly different association based entirely on their experiences, biases, and context. It's like trying to fake out your teacher by repeating what you read on Wikipedia: it may sound right, but it doesn't have any further meaning beyond that. What may work is having a recording from a person who used specific information to perform an action played back to another individual. This may allow some kind of context and application of the information to transfer as well, but even this approach could have problems. For example, if you seek to transfer, say, knowledge of kung fu from an expert into a novice, what would happen if the novice's body isn't at the same physical level as that of the expert? Or something even simpler, like the expert is right-handed but the novice is left-handed?

THE PLAYERS OF THE FUTURE OF NEUROPROSTHETICS

While medical science is currently leading the latest developments in marrying human and machine, it makes up only a small part of the research into neuronal interfaces with the goal of the enhancement of the human brain. The private sector is looking at the science as well, which isn't surprising, since economic forecasters have estimated that the value for the coming cognitive enhancement market will be in the tens of billions of dollars.

One of the most public figures pushing for a combination of human brains and computers is Tesla founder Elon Musk. His interest in technology comes from a rather frightening place: to put it bluntly, he doesn't want humanity to become AI's pet. Musk's idea is a neural lace that will be injected into the brain via a syringe and form a mesh that will integrate the technology with the tissue. From there, the computer-human hybrid has a fighting chance against a machine, in terms of intellect. If that sounds pretty out there, don't think that Musk isn't serious: he's formed a company called Neuralink to make it.

As Musk sees it, the development and growth of supersmart artificial intelligences may one day result in AI that has no need for us. The neural lace will allow humans to, at worst, keep a competitive edge against the machines, and at best, integrate completely with the AI to come. In addition, the neural lace could also input stimulation into circuits that begin age- or disease-related declines in activity, possibly returning the brain of a seventy-year-old to the activity level it had when it was thirty.

Also on the playing field is Facebook's Mark Zuckerberg, who has been recruiting neuroscientists and engineers for a project called Building 8, which involves brain-computer hybridization of some kind—or, as he called it at the 2017 Facebook Developer Conference, a "direct brain interface." Reportedly, the Facebook-backed interface will be noninvasive (as compared to Musk's neural lace) and will allow users to "type" by just thinking the words.

A third major player in the field is Bryan Johnson, who founded the company Kernel to develop a neuroprosthetic that will aid individuals suffering from neurological damage, as well as, one day, boost intelligence, memory, and other executive functions. Like Musk, Johnson sees achieving a human-computer blended brain as a necessity, as artificial intelligence's capabilities grow and broaden its reach.

While there are any number of smaller startups reaching for the same goals, and other countries undoubtedly have their own programs, the elephant in the room is DARPA. DARPA, which can at times seem as friendly as a favorite science teacher and at others as secretive as James Bond, has at least ten projects attempting to combine human brains and computers under way via its BRAIN Initiative, which kicked off in 2013.

DARPA has demonstrated positive results when it comes to enhancing memory retention using electrical stimulation, as well as increasing the pace and effectiveness at which soldiers learn, and has announced a program to investigate the creation of a memory prosthesis, as well as to create its own high-resolution implantable neural interface. Many of the organization's plans for brain research involve better prosthetics and helping individuals with brain injuries or neurological diseases as well.

If we look at the larger picture of neuroprostheses and brain enhancement, it feels like we're in the days shortly before the emergence of the telegraph or the internal combustion engine. With those early technologies, there were hiccups and false starts and assertions that neither would ever come to fruition, but once they emerged, the world was never the same again. Our world is potentially on the brink of a change just as monumental in the coming decades.

All of this points to one question about Rick: If he does have a computer in his brain, how "smart" is he really? He knows a million facts, always knows what to do, and knows the locations of any number of universes or star systems and types of alien species, all of which sounds like a database of information. Maybe anyone, given the right neural augmentation, could become Rick.

BOOSTING SNUFFLES TO SNOWBALL

Using technology to modify cognitive performance in humans, rats, and monkeys has been done, albeit experimentally. In every case, though, it's been assistance, not enhancement. With Snuffles in "Lawnmower Dog," it was pure enhancement.

The concept of increasing animal intelligence to near or above human levels is known as "uplifting," and it's a task that we've already achieved, as long as we use the very simple measure of intelligence of being able to perform cognitive tasks quicker and more efficiently.

"First off, we need to define intelligence," says neuroscientist Daniel Korostyshevsky. "Simplified, there are, in essence, two types of intelligence: crystallized—the ability to rely on previous experiences to solve problems—and fluid—the ability to solve novel problems. Both reside in distinct neural networks in the brain. It would be a moot point to be able to instill humanlike crystallized intelligence in an animal such as a dog. Self-awareness would be a prerequisite for such intelligence, and here it is obvious that Snuffles has it—he's able to recognize himself in the mirror in the episode.

"In the case of Snuffles, however, it's clear that he was relying on his doggie experiences in order to solve problems—that is, his fixation on his missing balls." The concern Snuffles—who renamed himself Snowball when his intelligence was augmented—showed for his missing bits is an example of crystallized intelligence,

Korostyshevsky explains, but fluid intelligence doesn't rely on pre-vious experience. For Snowball to have a human's level of fluid intelligence, there would have to be some rewriting done in the brain.

"Causing the correct neural networks to arise in order to demonstrate something similar to a human-level fluid intelligence could, in theory, be done a number of ways," Korostyshevsky says. "Electrical stimulation has in recent years been shown to be useful in directing neurons to make new, targeted connections in the spi-nal cord. This has been used in mice to repair spinal cord damage with moderately positive results. With the help of specific—but as yet not fully understood in the central nervous system—signaling molecules and targeted electrical pulses, it would, in theory, be possible to build these networks in a dog brain. In humans, this system is thought to reside in the frontoparietal network of the brain. For this to work with Snowball, we would need to assume that Rick has a vastly superior understanding of the brain and how consciousness and cognition arise. Whether this intelligence would be remotely similar to a human's is up for debate."

In addition to the issues raised by Korostyshevsky, there is the issue of anatomy—human brains are distinct, different, and special in the animal kingdom, so nothing should be able to be "turned on" and suddenly have human-level intelligence because the brains aren't the same. Similar, sure, but not a 100 percent match.

Of the similarities between the brains of dogs and humans, per-haps the most surprising is that dog brains have regions that are attuned to human voices. Monkeys, too, share this development. Recent research has shown that these regions can also make sense of the voice's emotional content. The finding isn't exactly surpris-ing if you own a dog, but still, it shows that human and canine brains share features at a fundamental level.

While human and dog brains share some similar functioning, there are obvious differences: dog brains devote 40 percent more processing power to making sense of smells than we do, and their auditory cortex can process sounds up to a frequency of 60 KHz, while ours tops out at a mere 20 KHz. But neither of these factors speak directly to what makes humans humans and dogs dogs. You have to look at other brain areas to drive the difference home—and for that, we need to talk about the neocortex.

The neocortex in humans is the largest part of the cerebral cortex, which is the outer layer of the cerebrum. It's the top layer of the wrinkly part of the brain, can be found only in mammals, and is the most developed part of our brains in terms of layers, neuronal organization, and types of tissue. This is the part of the brain that allows us to associate, generalize, build models and representations in our minds, and adapt to new situations and environments from similarities to those previously encountered. If you try to relate those behaviors and abilities to a dog, it's pretty clear that the neocortexes of dogs aren't as well developed as those of humans. That explains at least part of a dog's ability to be constantly surprised when it goes to a place it's been several times before, and why it's so excited to see you when you come home. It also partly explains why, if you want a dog to learn a trick, the behavior needs to be repeated many, many times in as many different locations as possible. This isn't to say that dogs don't have memory—there's plenty of research and anecdotes that suggest they do—they just don't have human memory.

But let's not get overconfident: we're not the only things on the planet with well-developed neocortexes. Dolphins and apes are widely known to be "smart." Using a basic standard of brain size related to body size as an indicator of intelligence, dolphins come in right under us, with chimpanzees, our closest primate relatives, below them. While the ratio of body size to brain size is a general indicator of intelligence, again, the cerebral cortex and neocortex are where it's at.

Research has shown that dolphins possess self-awareness, have their own distinct personalities, possess memories, can problem solve, can teach others, and can think about the future. Likewise, primates such as gorillas, chimpanzees, bonobos, and orangutans show similar mental functioning. Elephants are proven to possess mental and cognitive abilities similar to those of dolphins, apes, and humans, and have been observed problem-solving, showing altruism, using tools, and expressing grief during what's been described as a "death ritual"—something that was known to happen only with humans and, previously, Neanderthals.

Rounding out a sampling of intelligent animals, pigs, crows, and other birds show basic but still present intelligence—levels of mental activity and ability once associated only with humans.

Of the menagerie above, primates, dolphins, and elephants appear to have well-developed cerebral cortexes and neocortexes. Learning about the brains of dolphins, primates, and elephants has been a double-edged sword. True, we learn more and increase knowledge, but this comes with the realization that these animals, which have been mistreated, hunted, or worse by humans for thousands of years, may be as intelligent as we are, just in a different way.

MECHANISMS OF UPLIFT

As mentioned previously, mice and monkeys have been "uplifted," in that they performed better at activities that were, essentially, "mouse" or "monkey" in nature. The invasive technology just helped to make them "smarter" mice and monkeys. So there is that.

Rick's method was noninvasive and involved a helmet that read Snuffles's thoughts and translated them for his human audience. Given that we're barely able to read the simplest of human thoughts via technology—either through invasive or noninvasive means—it's worth noting that this method is years and years away (if it ever comes to be).

If we're looking at the ways to uplift animals, maybe adjustments can be made earlier. With genetic engineering, we have already been able to alter genes and develop brains of embryonic (or earlier) animals in attempts to adjust their intelligence.

As reported in 2014, Dr. Steve Goldman of the University of Rochester Medical Center in New York injected human glial cells into the developing brains of mouse pups in a location where they developed into a specific type of glial cell called an astrocyte. Glial cells are native to the brain and serve to support the neurons. Over the course of a year, the human glial cells overtook the mouse glial cells, to the point that you could legitimately call the mouse's brain half-mouse and half-human. It's important to note: this wasn't really a science-fictional thing, as lurid headlines like "Scientists Grow Human Brain in Tiny Rodent!" (*Daily Express*, April 19, 2019) may suggest. The human cells were there to support the neurons of the mouse's brain.

Astrocytes support neurons by strengthening the synapses between neurons, and a single astrocyte will have many tendrils reaching throughout the maze of neurons inside a brain. Human astrocytes, though, have more tendrils than mouse astrocytes.

The mice with the human astrocytes performed much better on standard memory and cognition tests than the mice with unaltered mouse astrocytes. By one measure, the memories of the adjusted mice were four times better than those of the unaltered mice.

Around the same time that human astrocytes were growing in the brains of mice, another study at MIT and collaborating European universities genetically altered mice to produce the human version of the Foxp2 gene, a gene associated with our capacity to learn and process speech. When the mice with the human form of the Foxp2 gene were tested for memory abilities in a standard maze activity, they were able to learn the correct route faster than their non-human-gene-carrying pals.

Goldman's team chose to pump the brakes on their work when they considered using monkeys, citing ethical issues, and it's more than likely other research groups will as well. If you consider that oft-quoted fact that the DNA of humans and chimpanzees is around 98 percent identical, this is where the potential for human uplift fueled by genetic modification could take place. It's something that science fiction writer David Brin, who's written many novels centered on animal "uplifting," has pointed to as the possible means for uplift: that there are genes that are switched on in human development that are switched off in chimps. According to Brin, that's where the attempts will be made to uplift animals.

But the chances of one adjustment creating an ape like those in *Planet of the Apes* are low. Intelligence research tells us that intelligence itself is polygenic—housed on many genes. Adjusting just one or two probably won't do the job, and the trials and failures would likely be brutal, and might ultimately be untenable, given our empathy for our fellow creatures.

There are other compassionate arguments that must be considered as well. The uplifted animals must be able to communicate, must be stimulated by their environment, and must be able to interact with their environment in ways that would support their intelligence. Perhaps uplifted animals that are "stuck" with no hands or no opposable thumbs would develop tools that would allow them to manipulate objects and their environment. Perhaps, like the dogs of *Rick and Morty*, they might build exoskeletons.

We also have to consider that animals that are "smart" might be doing the best with what they've got. The brain is a complex organ and requires a lot of support, both in development and throughout the life of an animal. Our brains use about 20 to 25 percent of the body's energy, but make up only about 2 percent of our overall body weight. Swedish researchers in 2013 produced guppies with larger-than-average brains through the use of selective breeding. The result was guppies with brains that were 9 percent larger than those of their fellow fish, and predictably, they performed better on cognitive tests (well, guppy cognitive tests). But the larger brains came with a cost: although the genetically modified fish had larger brains, the selective breeding had sacrificed their development in other areas, resulting in smaller guts and the production of fewer baby guppies. Uplifting animals so that their brains are more developed carries all manner of unintended consequences.

UPLIFTING ALIENS

As uplifting has moved from science fiction to reality, one of the questions that has been raised is "What (or maybe who) will these uplifted animals be?" As best we can understand, revenge is a human trait that's not shared by the animal kingdom. After all, think of the thought process that leads to the concept of revenge: holding a long-term memory that can evoke emotion, the ability to plan (whether that's in the short term or the long term), an understanding of the risks involved (which would start with self-awareness), an understanding that injuring or killing one of the pack may be for the best of the pack as a whole, and an ability to feel a semblance of satisfaction at the revenge that would limit its scope to just the single target. And of course, Snowball also understood the idea of a force multiplier, by building and outfitting other dogs with exoskeletons.

Animals with more developed cerebral cortexes, as mentioned above, may have some of those pieces, and some behaviors of those animals can possibly be interpreted as revenge, but it's not clear at all if animals with artificially developed cerebral cortexes possess all of them.

As Korostyshevsky sees it, Snowball's behavior was something of a wash in regard to what would be expected of a dog that suddenly possesses human-level intelligence but is still, ultimately, a dog. "Morality, altruism, and even the sense of fairness has been observed in a number of mammals," Korosty-shevsky says. "Wolves, for example, exhibit altruistic behaviors toward both other wolves and humans who have been living in close proximity to them. Chimpanzees are clearly outraged when they feel that they have been cheated. Empathy, in the form of wanting to help a fellow mammal in distress, has also been shown in wolves and many other mammals. If anything, Snowball was too cold and vengeful in light of ethological evidence. He would have wanted to protect his old 'pack,' castration notwithstanding."

UPLIFTING ETHICS

If all the above discussion of uplifting has led you to some uncomfortable ethical and moral questions, take heart; you're not alone. There are very passionate and reasoned arguments on both sides of the issue, and many countries are weighing in and drafting strict rules regarding experiments that would use animals that contain human material.

On the pro side, the views stem from the belief that if we can, then we should, even to the most extreme. These voices see uplifting animals as a moral imperative, in order to make the lives of animals better. Unsurprisingly, Brin is on the pro side of uplifting, seeing it as the right thing to do to help animals along and lend them a hand, to break them out of their brutal, survival-of-the-fittest way of life.

Those on the con side have multiple arguments, including the aforementioned sacrifices we'd have to make (which is to say, that the animals would have to make) until we perfect uplifting (and also whether we would even tell the first uplifted animal about all the dead creatures that came before it . . . er, them). Another argument against uplifting comes from the point of view that uplifting, at its heart, is all about human ego, and has nothing to do with any kind of benevolence. It's hubris in the extreme to suggest that animals would somehow be "better off" if they were more like humans.

That's not to mention the upset in the ecosystems and social structures of the Earth. Modern humans have a pretty lousy track record dating back tens of thousands of years when it comes to how we treat other species. After all, any story about uplifting, including "Lawnmower Dog," always has ominous tones—they're not happy tales in which everyone lives happily ever after.

Maybe we should worry about uplifting ourselves first.

THE FLIPSIDE: CAN YOU BE MADE TEMPORARILY DUMBER?

Yes.

When it was revealed that Rick had too many cybernetic augmentations to be allowed to fly commercial in "The Whirly Dirly Conspiracy," he was immediately dosed with a synaptic dampener that blocked violent tendencies and controversial thought. The drug wore off in six hours, allowing him to fly without posing a threat to the safety of the other passengers.

It seems unnecessary to point it out, but such drugs do currently exist in our world, too. In the monkey research mentioned earlier, which sought to enhance cognition, part of the experiments required that the learning and memory functions in the monkeys be, in essence, dulled or almost turned off. To do so, they used cocaine, which blunts the executive functions of the brain involved with decision-making, working memory, inhibitory control, and cognitive inhibition.

In our world today, the choices to dull intelligence and executive functioning of the brain are vast. To name a few: the active ingredient in marijuana, tetrahydrocannabinol (THC), has been shown to impact short-term memory, attention, and learning; psilocybin, the active ingredient in so-called magic mushrooms, can interfere with the frontal cortex, limiting complex thought; and opioids can produce a blissful feeling of euphoria that precludes any higher functioning in most individuals. Over thousands of years, humans have sought out chemical agents to alter our respective states of consciousness, often with the side effect of calming or shutting off certain functions of the brain.

And there are more focused ways as well, as Korostyshevsky says. "A number of years ago, researchers were able to shut down consciousness by using focused magnetic pulses to dampen specific parts of the brain. If you're looking to shut down intelligence—and here you would really need to specify what kind of intelligence you're talking about—you would in theory be able to do so with a pill. Just put a targeting molecule onto the active ingredient to either slow down or diminish the neuronal communications in the area of the brain that represents that specific intelligence, and it's done. A targeted meth-like molecule, for example, would slow down reuptake of neurotransmitters, making them stay in the synapse longer and prolonging a sensation, or, in this case, making thoughts linger longer."

The takeaway: making you smarter is a problem with many different approaches, some of them complicated to the point of brain surgery. Making you dumber is easy.

CHAPTER 12

Body Hacking

★ ★ ✦ ★ ★

Throughout the series, we've learned that Rick has the following cybernetic augmentations:

* Nanofiber defense mesh in his epidermis
* A cybernetic arm (shown in "The Whirly Dirly Conspiracy" and bitten off and regrown in "The ABCs of Beth")
* A cybernetic eye (that's good enough to see the president's Invisi-Troopers in "The Rickchurian Mortydate" and to be used as a targeting device in "Whirly Dirly")
* Additional enhancements/augmentations in his legs, heart, and brain (as seen in the security scan in "The Whirly Dirly Conspiracy")

And that's not even counting the enhancements we've seen on other characters, like those of the Vindicators' Crocubot, Snuffles's enhancements, the eye patch that Evil Morty wears, or Birdperson's change into Phoenixperson.

Oh, and Morty can turn into a car.

Okay, that last one is just . . . weird, even by *Rick and Morty* standards.

There are a lot of examples of technology making humans (and others) better in the series, but let's just focus on Rick. He's got enhanced everything: limbs, eyes, heart, brain . . . even skin.

For some, this might be a Frankenstein-like aspect of Rick—a scientist who can't leave well enough alone and who has the utter hubris to think that he can improve upon nature. But if he can improve it, why wouldn't he? After all, the human body isn't as strong as it could be, its materials wear out, and its operating system (nervous, muscular, circulatory, and other systems) could do with some upgrades, given what we've learned about computers, technology, and material science over the last few decades.

If we're going to discuss body hacking, though, we need to explore some vocabulary.

* Cybernetics: A broadly defined term that roughly means the marriage of biology and technology, with one critical particularity: the system must include a closed signaling loop.

* Closed Signaling Loop: This is an idea within cybernetics that the technological part is not just slapped on but is integrated into the overall biological system. That is, a change is caused by the system, feedback is given, and the system then changes. In everyday life, we experience this thousands of times. We reach toward an object, feel that it is hot, and move our hand away. It's so simple that our brains have evolved to not even waste time bringing it to the attention of the consciousness.

* Bionics: The popular term for the science of replacing part of a living organism with a robotic device, as well as for the robotic part itself.

* Cyborg: Short for "cybernetic organism," this term has been around since 1960, having been coined by Manfred Clynes and Nathan S. Kline when the two scientists were looking for a word to describe an enhanced

human being that could survive in extraterrestrial and other hostile environments. Again, it's a word with a very broad set of definitions depending upon who you ask; a person with an artificial pacemaker or even contact lenses could be considered a cyborg. A person using an exoskeleton in industry could also qualify, or it may refer to a blend of human and machine, packaged to look like a human or some mix of both, but the gold standard of "cyborg-ness" is that closed loop. With that, machine and human are the same, as the technology is recognized as part of the self.

And finally . . .

* Transhumanism: This is also a rather broad and flexible term, but it largely focuses on the idea that we, as humans, should increase our cognitive and physical potential through the use of technology. Though he is a scientist for sure, Rick's augmentations are transhumanist in nature, as opposed to being replacements for worn-out parts.

To put the terms together, you could say that, due to the increasing quality of cybernetics, we will transform into a society of transhuman cyborgs with more and more bionic parts, and you would not be wrong. It will undoubtedly strike some as creepy, but that doesn't mean it's wrong.

The fantastical idea of blending machines and biological life has been around for over a century. It originated in France in 1911, with the publication of the novel *Le Mystère des XV* by Jean de la Hire. The hero of the story was the Nyctalope, Leo Saint-Clair, arguably the first pop-culture superhero. The Nyctalope had an artificial heart, night vision, and other powers, all given to him by technology. Since then, there have been hundreds, if not thousands, of variations on the theme, to the point that the concept of a cyborg is widely accepted and understood to be something that blends biological life with computers or other technology.

Yet if we look at the whole picture, we see that this is another case of something that seems simple in fiction but is quite complex when we try to bring it into the real world. Still, advances in technology and science are rapidly bringing what was once fiction closer to reality.

It's worth noting that while Rick's technological improvements (as far as we know) are augmentations (that is, making his mind and body better than a human mind and body), that's not where the bulk of research is happening today. Existing cybernetics research is therapeutic in nature: focused on healing or mediating medical conditions and improving the quality of, and, in many cases, prolonging, life. While these therapies can produce replacements and enhancements that are better than the originals, improvement isn't the main goal. Those people who, like Rick, embrace cybernetics, who seek to enhance their bodies and senses with no underlying illness or condition as motivation, are the hobbyists, the grinders, and the DIY hackers.

MIND MEETS MACHINE—LITERALLY

If you're looking to control a machine with your brain (which includes your nervous system), whether it's your cybernetic arm, leg, eye, or anything that's controlled unconsciously, you need a way to communicate to the machine. That's accomplished via a brain-computer interface. While there are different types of BCIs, the goal is the same: to take information from the brain, a biological signal transmitted across neuronal cells via electricity, and turn it into a technological command that can then be sent either to muscles or cybernetic prostheses. Ideally, the prosthetic or device also sends a signal back, resulting in the closed loop mentioned earlier.

For example, when using a BCI with an advanced robotic arm in a case with a patient who has suffered brain or spinal damage, the command to pick up a mug of coffee would be generated in the brain of the patient; sent to the motor cortex, which initiates voluntary muscular movement; and then sent down the spinal cord, ultimately headed for the muscles that control the motor movement. Somewhere along that chain, depending upon what part of the brain is damaged, the signal would be captured by the BCI and relayed to the robotic arm, the processor in the arm would translate the signal, and the hand of the arm would perform the necessary functions for grasping the mug and picking up the coffee.

In theory, this seems quite simple. For amputees, the nerves that innervated, or provided the nerve impulses to, the missing limb are still there, so if you stimulate the nerve that would normally innervate the right index finger of a patient who has lost their arm at the shoulder, the patient still "feels" a sensation on their right finger. This is where the phenomenon of "phantom limbs"—the feeling that amputated or otherwise missing limbs are still there—comes from. In this instance, to move the patient's index finger, all the BCI must do is grab the signals from the brain, send them to a device that follows the brain's wishes, gather input from the device, and send it back to the brain.

Actually, though, doing this is not easy. Your brain has about 100 billion neurons in it, with about a million more in your spinal cord. Pulling a clear electrical signal from 100 billion chatty nerve cells isn't a simple task. Different parts of the brain do different things, and not all nerve cells are talking at once, but still it's not easy.

BCIs come in three varieties: invasive, partially invasive, and noninvasive. Invasive BCIs are exactly what they sound like: electrodes implanted directly into the gray matter of the brain. The electrodes can then be accessed via a port on the skull. Given that the brain is divided into different regions, each of which controls certain functions, the electrodes are placed in the region of the brain where assistance is needed, such as the ocular cortex for vision or the motor cortex for movement. Invasive BCIs were first used to assist patients with their vision, but have since been used to assist individuals with movement issues, whether to restore movement or allow them to move a robotic device.

Partially invasive BCIs are implanted inside the skull but are generally not sunk into the brain's gray matter. These interfaces show less risk of complications and rejection than invasive BCIs, and tend to require less training than those that are fully invasive. As with all BCIs, the output is fed to the device.

Noninvasive BCIs often focus on controlling devices via electroencephalography—that is, reading brain and nervous system activity through sensors on the scalp or other locations. As a technology, noninvasive control of devices is low-cost and comes without the risks and complications of surgery, but the skull can dampen the signals coming through. A way around these dampened signals, at least in some applications, is noninvasive BCIs developed for brain control of prosthetic devices that read nerve signals at the site of the amputated limb. If the amputation didn't involve nerve damage, the nerves are still present at the site of the amputation, and still fire when the individual thinks about the limb that the nerves once innervated.

A step up from that is targeted muscle reinnervation. In this approach, the nerves that originally continued to the limb—we'll use a hand as our example—can be reattached to the chest or bicep muscles of the individual. The nerves grow into the muscles, and as a result, specific regions of the muscles respond when the patient thinks about moving the limb. This muscle action acts to amplify the nerve signal, allowing the signals to be recorded through the skin, again keeping things noninvasive (after the initial reinnervation). The nerve locations are mapped and noted, so the researcher and patient will know what muscle contraction corresponds to which finger being moved, and those contraction patterns are programmed into the prosthetic.

Now, when the patient thinks of moving their finger, a portion of the chest or bicep muscle contracts; that signal is read by the interface, and the corresponding instruction is sent to the prosthetic.

With this approach, there's virtually no training involved. All the patient has to do is think about operating their hand as they used to, the rewired hand nerves cause a portion of the rewired muscle to contract, and that signal is read and sent to the bionic limb. And, with appropriate reinnervation and sensors, patients can "feel" sensations in their missing limbs in the same manner, as signals are sent in the opposite direction, from the prosthetic to the nerve via the interface.

Additionally, microprocessors in the limbs themselves can learn more complex movements as the patient thinks about them, which makes the newest mind-controlled prosthetics both natural-looking and a little uncanny. The arm "learns" how you want it to work, rather than your having to learn how the replacement arm works.

Given the augmentation to Rick's brain that was highlighted in "The Whirly Dirly Conspiracy," he may have performed neurosurgery on himself to allow him to control his augmentations, although when it comes to arms and legs, the noninvasive interfaces allow for near-constant updating and tweaking without further surgery.

GOING OUT ON A LIMB

We may never know how Rick lost his arm, but he seems to have gotten over it with a replacement that looks completely human—

um, except when it opens up into a high-powered gun. It's also a simple-enough setup that he carries spares just in case he loses the arm again—by, say, its being bitten off in a creepy fantasy land.

While Rick's bionic arm is indistinguishable from a regular arm, his replacement looked similar to the most current versions available today. The Modular Prosthetic Limb (MPL) is just one example of advanced arm prostheses. Developed by Johns Hopkins's Applied Physics Lab in association with DARPA, the arm has twenty-six degrees of freedom, seventeen motors with onboard controllers and sensor conditioning, and one hundred different sensors that can send signals back to the brain, which creates a rudimentary sense of touch and feel. And it looks like it came straight out of science fiction.

According to the patients who use the MPL, using it feels like you have a hand. And what makes it even cooler is that the connection between the individual and the arm is wireless rather than hardwired in. That means that an individual using the MPL can use it from across the room. Maybe that's more creepy than cool. You know those science fiction images of a crawling robotic hand? That's now totally possible.

In the case of individuals who have suffered spinal damage that prevents any bodily motion, invasive interfaces have been developed with their own bionic arms that can be controlled by the brain of the paralyzed person. Patients using the more advanced models being developed at the University of Pittsburgh can remotely control and receive sensations from a robotic hand directly through the interface, which connects to a port mounted on their skull.

Given the need for working hands, bionic arms tend to get the most attention in the news, but research on new bionic legs is just as cutting-edge. MIT professor and double-leg amputee Hugh Herr is famous for his two bionic legs, attached from the knees down. Herr, who's also a mountain climber, lost his legs in a climbing accident in 1982, and quickly started to adapt the prosthetics available at the time in order to return to climbing, and even to climb better.

In their current form, Herr's legs allow him to walk at different speeds, mimicking the action of many muscles in the legs. His research stresses "biological mimicry," which emphasizes the use of "smart" prosthetics that follow the design and action of actual legs. Among the benefits is that there's a very small learning curve for patients, because their bodies remember how to walk. And that's the thing: watch a video of Herr walking, and he looks perfectly normal. If he didn't have his pants tailored to show them off, you wouldn't know that he had two bionic legs.

Herr is one of the researchers leading the way toward humans becoming cyborgs. He laments in his TED Talk that he's "just a bionic man" rather than a full cyborg—that is, his bionics don't give him the true sensation of touch and feel.

And just to show that researchers in the field aren't without a sense of nerdery, one of the most advanced bionic hands, which does allow wearers a sensation of touch, is called the LUKE Arm, named after famous hand amputee Luke Skywalker.

SIGHT, TARGETING, AND MORE—BIONIC EYES

Rick revealed his bionic eye when it showed up in "The Whirly Dirly Conspiracy" to aid in targeting the crime boss after the trip through the wormhole, but it's been there all along. We can admit it: his bionic eye is better than our biological ones.

Our eyes are marvels of evolution, but they have their limitations: we can see only a teeny-tiny portion of the full electromagnetic spectrum, our ability to see close up and far away is not terrific, the range of colors we can see in the visible-light portion of the spectrum is only a part of the colors that exist, our eyes are made of vulnerable material that can easily fall apart or cloud over, the range of brightness in which they work is pretty narrow, and, frankly, they wear out waaay too quickly—after only forty years, they tend to diminish significantly. What kind of design work is this?

But again, current eye augmentation isn't aimed at turning our eyes into glowing cameras capable of seeing everything. Like work with bionic limbs, it's aimed at helping those with visual impairments of many varieties.

In terms of technological improvements to the eyes, there are two regions to work on: the front, where the light comes into the eye, and the back, where the pattern of light is registered and turned into nerve impulses. We're already well into using tech to fix eyes, with laser surgery and corneal replacement. Move to the back of the eye, and now you're talking about the development of retinal implants that can "read" the signals (images) coming in through the eye and "write" those signals to the retina if it just needs help, or to the optic nerve directly. When the retinal implant (often called a "bionic eye") stimulates the nerve, it causes the person to see phosphenes—small light bursts without any light actually entering the eye. The phosphenes allow the person to map their surroundings to the point of being able to see separate objects on a table. Resolution, however, is low.

Unfortunately, we're years away from having a completely replaceable human eye analogue that can be slapped into an empty socket, but the foundational work is being laid.

NO ONE HAS TO BE LIKE THE TIN MAN

Rick's heart was clearly labeled as something that has had augmentation, or just might be completely artificial. And why not? Like our eyes, our hearts could be better.

Current cardiac technology can keep a heart beating via a pacemaker or allow for a total replacement with an artificial version that's most often used as a stand-in until a human donor heart can be found but can, in the best circumstances, completely replace a human heart.

In our reality, a permanent artificial heart for humans is just on the horizon: the first is already being developed at Oregon Health & Science University. The OHSU heart is simple in design, with one moving part: a hollow aluminum rod that moves back and forth in a titanium tube that acts like the ventricles of a normal heart and mimics normal blood flow, sending the blood first to the lungs and then to the rest of the body. The artificial heart also has a "pulse," which is different from prior models that just had a constant flow. Power is supplied by a battery pack that the individual must always carry and recharge as needed, and will likely one day be replaced by a subdermal battery pack.

Also in regard to artificial heart technology, there is a push toward creating customized artificial hearts using 3-D printing, which would either be entirely artificial or created in a cyborg form, with artificial components being combined with printed tissues generated from some of the individual's own cells.

Pacemakers, on the other hand, have been used for decades to electrically stimulate heart tissue to help hearts that otherwise have trouble maintaining a steady beat. They've also become much more advanced than their original versions. Modern pacemakers communicate via a wireless network (which can actually be hacked, as recent news stories have reported) and analyze the rate and detail of the heart's individual beats, which allows them to adjust to the patient's needs, such as increasing the heart rate during exercise. While pacemakers also require batteries, recent research is looking at outfitting the devices with piezoelectric ceramic, which can convert the squeezes and vibrations from a beating heart into electrical signals that will keep the pacemaker working as long as the heart keeps beating.

While artificial hearts and pacemakers fall into the bionic category, these technologies or augmentations aren't necessarily seen in the same light as artificial limbs. Like eyes, replacing or assisting hearts is virtually always therapeutic in nature, with the aim of improving or prolonging life. Artificial hearts may be a transhumanist target, but for now, the main goal is developing technology that assists or replaces a malfunctioning heart.

THE BIOHACKERS, THE GRINDERS, AND THE BODY MODDERS

Virtually all the modifications discussed in this chapter so far are done out of medical necessity, but some people are interested in bionics with other motivations.

There is growing subculture of transhumanist DIYers. They go by a few different names, but are all similar in that they see being "merely" human as something to get away from, something to "overcome," to get all Nietzschean about it. Most people can relate to some sense of dissatisfaction at some point in their life with their body, but these people aim to change those things with technology.

These individuals have been known to put any number of devices in their bodies: RFID (radio-frequency identification) chips for identification, magnetic slivers in fingers to "feel" magnetic fields, earbuds in the cartilage of the ear, subdermal magnets that allow for cosmetic attachments, and, unsurprisingly, implants aimed at increasing sexual pleasure. Biohacking has tended to be unregulated, but more and more, as companies look to tech for a variety of solutions from employee identification to patients who carry their medical history with them at all times on a subdermal chip, it finds itself on the cutting edge of innovation.

Hacking has moved from just making tweaks and novelty adjustments in the body to making individuals more enabled than they were before. And, in a pattern that mirrors the growth of the PC industry, the lines between the DIY crowd and the medical community are blurring, with the potential for innovation sky-high. Biohacking gatherings, which were once the domain of hobbyists, are now conferences attended by interested doctors and medical students. Start-ups may not carry the same cache in the business world that they once did, but new, legitimate biohacking companies are showing up every day, and they have the eyes of finance upon them.

WHERE DOES IT GO FROM HERE?

As a species, we have a good record of putting our abilities in front of what is best for society, so ethical questions are already an integral part of biohacking, and will likely remain so. Where do we draw the lines for what we can and cannot do with our bodies?

MIT's Herr is a full-on evangelist for a world of cyborgs, with his idea of bionics and cybernetics reaching far beyond limb replacement. In his widely viewed TED Talk, he considers a near-future filled with humans who possess enhanced strength due to reconfigured muscles and embedded motors, humans with wings, and humans with other nonanthropomorphic structures—a world filled with humans who are unrecognizable to what we are today.

But it's a future that should be approached slowly and with deep consideration.

The changes would make humans unequal, which some argue would immediately create more stratification in society. Enhancements, as Rick demonstrates, make people stronger and faster already. By virtue of its materials and design, the MPL can curl forty-five pounds. Before his fall from grace, South African track star Oscar Pistorius was one of the fastest men in the world, partly due, as many argued, to an unfair advantage given to him by the twin blades that he wore on his legs as a double amputee. We may be creating a world of superheroes, as Herr suggests, but if we're not all superheroes, human nature has some problems baked in—like fearing things that are different or that we don't understand, as well as putting "different" people into categories and treating them differently.

And we should focus on the fact that in the United States, one of the major backers of advanced prosthetics research and design is the military. There is obviously a serious need for better care for soldiers wounded in battle, including those missing limbs due to combat, but DARPA also funds research into exoskeletons and other human-enhancing projects. An injury so grievous that a solider would have to have a limb amputated in the field used to mean an automatic return home to convalesce. In a few years, will the doctor just attach a bionic arm to the cauterized stump, make sure it works, and send the soldier back out?

How will we, as a society, react when someone publicly announces their plan to amputate a perfectly good limb and replace it with a bionic one that can do more? Will we riot in the streets, or will we all want guns built into our new arms, too?

CHAPTER 13

Cockroach Brains

★ ★ ⭐ ★ ★

Changing oneself into a pickle to avoid going to a family therapy appointment is one thing. Rolling into the wastewater runoff system under your town (which happens to be populated by cockroaches and rats), taking charge, and creating a nerve-tissue-controlled exoskeleton for yourself is something else altogether.

There are so many things to talk about in the "Pickle Rick" episode aside from Rick turning himself into a pickle, but it all starts with being able to move as a pickle; and to do that, Rick had to hijack a cockroach's nervous system . . . which is something that is totally possible.

That's right: this is one of those few places where *Rick and Morty* science is pretty much in line with real-world science—that is, when it comes to controlling cockroach brains. Turning oneself into a pickle is still a little out of reach.

In the episode, when Rick has fallen down into the wastewater drainpipe, he lures a cockroach closer by biting his own pickle lip and "bleeding" out some brine. Roaches (and these looked to be American cockroaches, good old *Periplaneta americana*) aren't necessarily attracted to brine and salty flavors, though. In fact, salt has long been reported as a homemade roach repellent (survival bunker tip #101: soak food packaging in a salt solution). But roaches are omnivores, and they're curious. They'll sniff with their antennae and otherwise probe all around an environment just to find food, and a novel smell—say, that of a pickle—more than enough to attract the roach. The puddle of brine just concentrated the scent.

Once the roach is close enough to his mouth, Pickle Rick chomps the junction of its head and thorax, killing or stunning it (it's hard to tell for sure). He gently removes the part of the roach's exoskeleton that covers its brain and pokes around with his tongue until the roach's legs start moving. From there, he manipulates the roach's legs to maneuver his pickle body on top of the roach and then controls the roach as it moves him along the ground. Later, he's shown to have made a system of roach legs that he controls via a series of tissues that he manipulates with his tongue.

Please don't go poking your tongue into cockroach brains to see if you can move the legs—it's not very hygienic, and you'll lose friends, no matter how cool you think it is. Besides, you're much too big. What Rick did might have been accurate when it comes to the real-life science of cockroach brains, but there's no need to practice it. If you're willing to use something other than your tongue, though, it's not only possible, it's pretty easy. So easy, in fact, that there's a company that will sell you a kit that teaches you how to do it for yourself.

FIRST THINGS FIRST: COCKROACH ANATOMY 101.

A cockroach is an insect, so its body is made up of three distinct regions: the head, the thorax, and the abdomen. Most roaches have only vestigial wings and either can't fly or can fly for only short periods. Roaches have antennae on their heads and two antennae-like sensory receptors called cerci at the back end of the

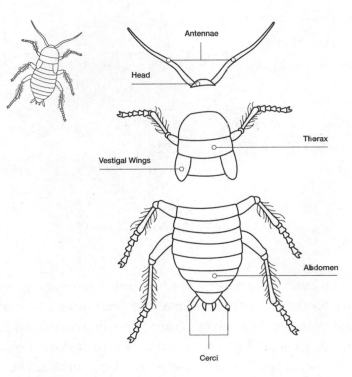

abdomen. Among the cockroach's eyes (which have a thousand lenses), its cerci, and its antennae, it can sense (and respond to) stimulation extremely quickly. Ever try to sneak up on a cockroach? You can't. It's like they know you're coming. Because they do.

Sensory-wise, as far as the cockroach is concerned, a human attempting to sneak up on it would be like a marching band playing "Nearer My God to Thee" at full volume. People give off smells, change the local temperature and humidity, and move the air in weird ways compared to when they're not there. We're just big noisy things to roaches. If we try to catch a roach, we might get lucky now and then, but more often than not, the roach gets away.

The nervous system of a cockroach is pretty simple, and that's

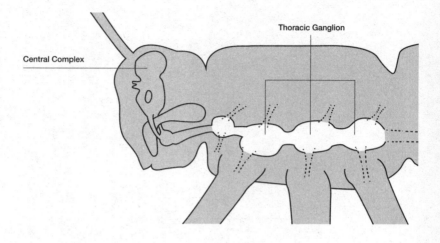

not an insult. It's a reference to its classification and organization. The cockroach nervous system isn't centralized like ours. The main parts are the brain (also called the supraesophageal ganglia) and the ganglia. If you look at the cockroach from the side, the brain would be above the esophagus. Also in the head but below the esophagus are the subesophageal ganglia, which connect to the double spinal cord that runs through the ventral side of the body and connects to six more ganglia.

Obviously, "ganglia" is the key word here. Simply put, they are clusters of nerve cells in one place, appearing as swellings along the nerve fiber. They're not quite "brains," since they lack more complex organization, but they are centers of nerve control and signaling. It's this distribution of the cockroach's nervous system that allows a headless roach to live for a week after decapitation, and even then, it only dies of thirst.

Back up to the brain and subesophageal ganglia: the roach "brain" is largely concerned with incoming sensory stimuli, while the subesophageal ganglia focus on motion—controlling the legs, the wings, and the mouth.

SECOND THINGS SECOND: NEURONS AND ACTION POTENTIALS

We learned before about how nerve cells, or neurons, fire, but let's hit it again with a little more detail.

The inside of a nerve cell is more negatively charged than the outside, thanks to the neuron actively pumping positively charged sodium ions out while keeping a few positively charged potassium ions and negatively charged protein molecules in. The inside has a value of about -70 mV compared to the outside, which has more positive ions. The value of -70 mV is called the cell's resting potential, and we can say that the interior of the cell is polarized—it's a region with its own distinct charge.

When the nerve cell receives a stimulus—say, in our roach's case, from an antenna—sodium channels in the cell membrane open, allowing some sodium ions from the outside in. Bringing positive ions into the neuron starts moving its charge toward the positive. If the charge moves to -55 mV, then voltage-sensitive sodium channels in the membrane open and sodium ions flood into the cell, depolarizing it. This depolarization fires the cell, sending a signal, called an action potential, down its branches (called axons) in a wave. The depolarization causes the axons to release a chemical signal (a neurotransmitter) of their own to the next nerve cell, which acts as the stimulus for that cell.

Meanwhile, back in the first neuron, following depolarization, potassium channels in the membrane open and pump positive potassium ions out, once again decreasing the charge inside the cell and making it more negative again. The charge in the cell can fall to around -77 mV (called hyperpolarization) before the neuron's sodium/potassium pumps work to return the cell to the resting potential of -70 mV. Things are now ready to go again when the next stimulus hits.

THIRD THINGS THIRD: THE CENTRAL COMPLEX

Pickle Rick poked a part of the roach's brain with his tongue—his salty tongue, loaded with sodium, potassium, and other ions that were ready to change the electrochemical composition of wherever they landed in the roach's brain. Precise anatomical location aside, there is a spot in the insect brain where, if you poke it, you'll get the legs to move (among other things): the central complex.

The central complex is an unpaired part of an insect's brain (whereas there are two of each other structure), and it can be thought of as the hardwired instructions for what to do in response to different stimuli received from any number of sources—tactile, visual, olfactory, or other senses. The central complex is essential for sensorimotor integration. The neurons located there encode the sensory information received by the antennae, cerci, and eyes, and then send out responses.

Key to our understanding of the functioning of the cockroach's central complex was an experiment performed by Joshua Martin and colleagues, published in *Current Biology* in 2015. Martin inserted tiny electrodes into cockroaches' central complexes and recorded the action potentials produced while the animals were freely exploring a habitat. What they found was that the bulk of the cells' activity directly correlated to the speed at which the roach was moving, and its angular velocity—that is, how often it turned from a straight path. Perhaps even more interesting was the fact that the neurons fired before the motion they ultimately controlled; to put it another way, the nerve activity predicted the roach's movements. The movements prompted by the signals from the central complex covered the space around the roach in its entirety.

The researchers realized that the central complex contained the sensory maps for the insect, which let it know what direction incoming stimuli were coming from and also constructed a complete map of the immediate directions in which the roach could move, along with the ability to initiate and continue that movement. Martin and his colleagues tested this by taking the signals they had recorded from the roaches' central complexes and sending them into another roach's central complex. Sure enough, upon the reception of the signals, the other roaches moved their limbs in the directions previous recording indicated, and at the same speed as well.

In short, if you want to think of a roach—and other insects, since the central complex shows up and serves a similar function in bug after bug after bug throughout evolution—as a tiny robot with preprogrammed instructions that is just waiting for the right neuron to fire and kick things off, here is your permission to do just that. Martin's group made their roach subjects walk at varying speeds and turn in various directions just by stimulating parts of the central complex. That part of the brain is located roughly in the center, more or less where Rick's tongue was poking. And it took him only a couple of tries to get those legs moving.

When Rick kills his first rat later in the episode, we see the extent of his cockroach handiwork. His whole factory that removes the rat's brain, skins its body, and attaches rat limbs and other parts to Rick's pickle body, is all made from roach limbs. Rick controls the movements of the factory's pieces by touching his tongue to the blob of roach brain just under his mouth. That's a central complex taken out of a brain and connected to nerves running throughout the factory—the nervous systems of dozens of roaches, connected and stretched to respond to stimuli from the main central complex that Rick controls.

That's not to say that anyone could make this for real, but in theory, it's possible. The roach legs would all have their preprogrammed instructions thanks to the central complex, and again, in theory, multiple roach parts could be "wired" together so that a stimulation flows from one set of nerves into the next set down, with ganglia spread throughout. To be fair, the trial and error Rick would have needed as he figured out the mechanics, levels, and push-pull combinations of roach legs would've made for some boring television, and in reality would've taken far longer than the hour-long therapy appointment Rick was trying to avoid. Also, the nerves would have to be tended to, hydrated, not overstretched, nourished, and have their signals amplified to flow for distances much longer than those inside a roach's body, but making a factory where you've essentially rewired a bunch of organic robots into the tools that you need—that's pretty cool, even for Rick.

NOW, YOU DO IT. KIND OF.

If all this has gotten your inner mad scientist going, you're in luck. Cockroaches have been a model for researchers studying autonomous robot systems for years. The goal is that someday (sooner rather than later), researchers and engineers will be able to create autonomous robots that mimic insects in their motion and control systems, or (even weirder) outfit cockroaches with computer chips that will allow their movements to be controlled.

The technology for remote-controlled roaches—or, to use a term that doesn't make people recoil at the idea of robot roaches, "biobots"—is relatively simple and can be created even by total amateurs. More on that a little bit later.

Although it is an option, the simplest way to control roaches isn't to go in through the central complex—it's too, well, complex. A control module that would interact with the central complex would need a lot of electrodes, and, as they say in insect brain labs around the world, "mo' 'trodes, mo' problems." Okay, they may not say that, but the problem is real: electrodes can get jostled if the connections between them and their chips become loose, possibly resulting in scrambled roach brains; the electrodes can become corroded and start producing oxygen and other ions due to the current moving through them (also bad for the roach brain); and initial placement can be pretty finicky. Controlling roaches from the central complex would give finer-grained control of movements, but if you don't mind rougher controls, there's a simpler way: the high-tech cockroach cowboy way.

To oversimplify things a great deal, the direction that horses travel is controlled by the reins, which are held by the rider. If the rider wants the horse to go right, they tug on the rein that's connected to the right side of the horse's bridle, and the horse turns in that direction. That's the basis of this simpler way to control cockroach motion (minus a tiny cowboy, which, while it would be cool, would make things much more complicated).

This neural-stimulation method is made up of two parts: a chip-based backpack with a power source and circuitry, and two electrodes. One end of the electrode is soldered onto the chip backpack, and the other is inserted into the cockroach's antenna, which has been clipped off, and secured with glue. If the roach receives a stimulation via the electrode in the right antenna, the signal is interpreted by the central complex as an obstruction or a predator in the way of forward motion on the right, so the roach turns to the left. Vice versa for the opposite side. A proof-of-concept experiment by Tahmid Latif and Alper Bozkurt at North Carolina State University allowed the roach controller to wirelessly "walk" the remote-controlled roach along a predrawn serpentine path. To keep the cockroach moving, the researchers placed electrodes on its cerci, the sensory organs at the back end of the abdomen.

Since this method is so simple in setup and control, Bozkurt said he envisions adding sensors on the roaches as well, allowing a number of them to move through an environment and collect data—or look for sounds, smells, or other signals of life, for example, at the scene of a disaster.

This approach to cockroach control is so simple that a kit is available that includes everything you need to create your own "RoboRoach," including a backpack with a wireless receiver that allows you to control your roach from an app on your phone. Swipe either to the right or left, and the circuit board in the backpack stimulates the antennae via the appropriate electrode. Oh, and appropriate cockroaches (the bigger, the better) are available via animal supply houses, so you don't have to catch your own roach to experiment on. Unless you want to.

One other, perhaps more advanced cyborg-roach creation came from researchers at Texas A&M University who developed a system that directly stimulated the front-most (prothoracic) ganglion rather than needing the signals to be processed through the brain. According to their findings, this allowed control of the cockroaches' locomotion with high repeatability.

One thing the researchers at Texas A&M noted was that they chose their method (stimulating the ganglia directly, rather than going through the antennae) to avoid the phenomenon of habituation—that is, after receiving artificial stimuli (electricity) for a period of time, the roach's nervous system learns to ignore it, and the response rate drops off a cliff.

That's one slight issue with the whole scenario of controlling multiple legs by tonguing the central complex. Those nerves and ganglia making up Rick's larger factory would most likely habituate quickly and just start ignoring the stimuli. But Pickle Rick quickly yanked off and discarded a roach leg when it wasn't sawing well, so the idea that he knew about habituation (and had a stock of roach legs and ganglia) certainly isn't that far-fetched.

If you ever find yourself inclined to give this a try yourself, a company called Backyard Brains, based in Ann Arbor, Michigan, will sell you the kit and give you all the instructions you need to make your own app-controlled roach. Be warned, though: the process is not for the squeamish. Minor surgery on cockroaches is involved. You don't have to bite their exoskeleton from their brain and smash your tongue in there, but it does take steady hands and a strong stomach.

NO COCKROACH ARMIES

While some researchers are looking at the control of cockroach nervous systems to create swarms of biobots, world domination isn't the only goal of cockroach-brain hackers. Controlling cockroaches, not so much. But creating small robots that mimic cockroaches in both body and brain . . . now you're talking. Researchers at the University of California, Berkeley, created the CRAM robot based on a cockroach design and behaviors and claim that small, autonomous, virtually indestructible robots such as theirs would be ideal for search-and-rescue missions through rubble and confined spaces, among other things.

Cockroaches may creep you out now, but someday, a robot one just might save your life.

CHAPTER 14

Gravity

★ ★ ✦ ★ ✦

Let's start with some basics on gravity.

FUNCTIONAL GRAVITY 101

There's an old joke that says there's no such thing as gravity—the Earth just sucks.

It's an old joke, not a funny joke. But the joke gets across the idea that gravity pulls down, which, from our perspective as monkeys just out of the tree, has the same result as sucking us down would. But if we're going to be precise—and we are—then we need to define gravity as a force. It's the force that's always present between any two objects with mass, attracting them together. It's what causes an apple to fall from a tree and a planet to orbit a star, and it's responsible for the sensation we call "weight." The more massive an object is, the greater its gravitational pull on others.

Gravity is a force, one of the four fundamental forces of the universe. This force exists between any two bodies in the universe, no matter their distance from each other. Gravitational force is directly proportional to the product of the two masses and inversely proportional to the square of the distance between the respective centers of mass of the two objects. Gravity follows what's called the inverse-square law, which means if you double the mass, the

gravitational force doubles, and if you double the distance between the objects, then the gravitational force is one quarter of
its original strength.

Aristotle was the first person to write a quantitative description
of an object as it falls, but while various ancient scientists had
more or less gotten down the basics of gravity, we look at Galileo
as the first one to really grasp the larger picture. Galileo was the
first person to explain that falling objects increased in speed at a
constant rate, which we call constant acceleration. He also realized, through experimentation (which in and of itself was a revo-
lutionary development in the history of science), that the mass of
the falling object does not affect the speed of the fall. Ignoring air
resistance, all objects fall at the same rate—that is, with the same
acceleration.

It sounded as weird then as it does now, but try it yourself. Hold
something heavy and something lighter side by side, the same
height from the floor. Drop them at exactly the same time and see
which one hits the ground first.

Neat, huh?

The rate at which two things fall is 9.81 meters per second per
second (which can be abbreviated as m/s^2). In other words, for
each second of free fall, an object gains 9.81 meters per second
of velocity. Everything on Earth falls at that rate (again, ignoring
air resistance; in fact, from here on out, just assume I want you
to ignore air resistance unless I tell you not to). You could argue
that a flat sheet of paper falls slower than a pen because of the air
pushing against it, but crumple up the piece of paper and you can
see constant acceleration at work.

Galileo also did experiments on motion. He figured out that
if you fired a cannon from high up on a wall and the cannonball was traveling at 50 meters per second, it would travel at a
constant 50 m/s horizontally every second, but it would drop
5 meters vertically in the first second, 20 meters in two seconds,

45 meters in three seconds, and on and on. This application of science to projectile motion was the basis for Galileo's invention of the "sector" in the late 1500s, a compass-like instrument that allowed gunners to better calculate their aim and how much powder they needed to get a cannonball to the target.

A century later, Isaac Newton did a lot of thinking about the work of Galileo and others. Extrapolating Galileo's cannon idea to the extreme, Newton asked what would happen if a cannon was on a tower so tall that it was above the atmosphere, and the cannonball was fired really, really fast. The cannonball would still fall 5 meters down in the first second, but, if the cannonball was moving fast enough, the Earth would curve away below the horizontal. Fire the cannonball at the right speed, Newton theorized, and it would never lose any height; it would keep falling constantly. In other words, it would become a satellite orbiting the Earth. Using appropriate math, we can calculate the speed at which the cannonball must be fired so that it falls forever for any altitude.

Following along these lines of thought, Newton looked to the moon and realized that it circled the Earth due to gravity, and the gravitational force between the Earth and the moon decreased with the inverse of the square of the distance between the two bodies.

Combining what he had realized about gravitational force with elements from what would later be known as his Third Law of Motion, Newton put it all into one formula, the Law of Universal Gravitation, which looks like this:

$$F_g = G\,(m_1 m_2)/r^2$$

For this formula:

* F_g is the gravitational force, measured in a unit of force called a Newton

* m_1 and m_2 are the masses (not weights) of the two objects, measured in kilograms
* r is the distance between the two objects, measured in meters
* G is the universal gravitational constant, or Cavendish constant, which is always 6.67×10^{-11} $m^3/kg \times s^2$. (The units refer to the force, measured in Newtons; the distance, measured in meters; and the mass, measured in kilograms. Ultimately, they cancel out to leave you with the unit you need, depending upon what values you put into the formula.)

Additionally, we can calculate the strength of the gravitational field of any planet with the formula $g = Gm/r^2$

* g is the gravitational field strength, in m/s^2
* m is the mass of the planet in question, in kilograms
* r is the distance between the two objects, measured in meters

Solve this second equation for g using Earth's mass and radius and you'll get approximately 9.81 m/s^2.

Though the value of the Cavendish constant wasn't calculated until the late 1700s, and you need to use heavy-duty calculus to show how it works with planetary orbits, Newton's Law of Universal Gravitation was so spot-on that it's still used today as the basis for calculating the flights of probes through the solar system, as well as determining the forces acting on objects throughout the universe. But (and here's a spoiler for what's coming up) though his math was good and his formulas can still be used, Newton was fundamentally wrong about the nature of gravity. It would take our old friend Albert to set things right.

WAIT . . . ABOUT WEIGHT

A good understanding of gravity requires us to separate the concepts of "weight" and "mass." They're not the same.

Mass is how much stuff (atoms and molecules) is in a given object. It never changes for that object (unless you change the object itself, but then you won't really have the object you started with anyway). In science terms, it's measured in kilograms. As for why it's a kilogram and not the prefix-free gram, as with the other units we've seen, that's a story from history. By the command of Louis XVI, a new measuring system—which would later evolve into the metric system—was created that decreed that the base unit of mass would be the mass of a liter of water at zero degrees Celsius, which is about 1 kilogram.

Postrevolution, the new Republican government of France kept the fledgling metric system but changed the base unit of mass from the kilogram to the gram, since many masses that needed to be measured in daily life were smaller than the larger unit. But despite the need for it, establishing a standard gram mass was quite difficult at the time. As a result, the decision was made to go back to the more massive (and easier to create and manipulate) kilogram, clunky prefix and all.

Weight is the interaction between the local gravitational field and the mass of an object, expressed as the product of the mass and the acceleration due to gravity. As a force, weight is measured in Newtons.

This relationship can be expressed mathematically like this:

$$F_g = mg$$

* F_g is the gravitational force, aka weight, measured in Newtons
* m is the mass, measured in kilograms

* g is the acceleration due to gravity of your local region; for Earth, that's 9.81 m/s²

For example, the weight of an object on Earth that has a mass of 1 kilogram is 9.81 N.

If an object has more mass, there's more of it for gravity to pull on, and therefore it has more weight. Less mass, less weight. Look at a physicist the next time you say you want to lose weight and you'll see a small crinkle in their nose indicating distaste, albeit fleetingly. You want to lose mass. If you want to change your weight, you need to move to a place where gravity is either stronger or weaker than it is in your present location.

As Rick, Morty, and company move from planet to planet, their respective weights should change. They would weigh less on a smaller planet that has less mass, such as Dwarf Terrace 9 (Tiny Earth), and more on a larger planet that has more mass. Bring biology, physiology, and anatomy into this mix and weird stuff starts to happen. With less gravity acting on the anatomy and biology you grew up with on Earth, you'd be able to jump higher, lift more, and do other crazy things, because you'd find yourself to be "stronger" than you are back on Earth.

Think of the footage of Apollo astronauts on the moon, skipping, hopping, and jumping. It looks like they're in slow motion, but they're not. The moon has one-sixth of the gravity of Earth (the approximate value of g on the moon is 1.65 m/s²), so all things, including astronauts, fall slower there than they do on Earth. Similar effects will be seen when humans send back video from Mars (Matt Damon in *The Martian* notwithstanding)— Mars's g is 3.72 m/s², about 38 percent of Earth's. For the Martian explorers, some adjustments will be needed, and for viewers of the video on Earth, there will be something uncanny about it that many people won't be able to put their finger on until the difference in gravity is clearly explained. For observers, objects will take longer to reach the ground when they fall, explorers' gaits will be different from those of people on Earth, and their jumps will be just a little higher. But hey, on the upside, if you live on Mars, you weigh less!

Weight and apparent strength are the fun side of living in less gravity, but there would be drawbacks too, and they're all internal. As long-term studies from the International Space Station (ISS) have reinforced, living with reduced gravity takes a toll on our bodies and our health. Muscles that aren't used as they were on Earth will atrophy, which is why astronauts must work out with resistance training every day, otherwise they'd be as weak as kittens when they came home. Balance in a gravitational field, which doesn't exist on the ISS or during a flight to a new planet, is also compromised, and needs to be reoriented. Osteoporosis (that is, reduction of bone mass) kicks in, since bones aren't stressed by supporting the body's weight as they normally are. In addition to this, there are vision problems due to pressure changes in the eyes, cognitive ability decreases slightly over longer periods of time, and there is evidence of alterations in gene regulation. Basically, any system that depends upon gravity's pull for feedback or monitoring is going to show some changes. It turns out that we're just bags of jelly on legs after all.

(An aside: Since we're talking about gravity, we should say that astronauts on the ISS are not in "zero G" or zero gravity. Yes, they look like they are, and in reference to the ISS that surrounds them, they are floating freely—but it's not because there's no gravity. The ISS is only 400 kilometers up. Earth's gravitational pull is a little less than it is on the planet's surface, but it's still the boss. After all, there's enough gravity to keep the ISS in orbit, which means the astronauts are also still experiencing gravity. You can even figure out how much using Newton's Law of Universal Gravitation if you'd like. Just remember to add the ISS's altitude to the radius of the

Earth and you're good. The masses would be the mass of the individual astronaut you're concerned with and the mass of the Earth.

What the astronauts are experiencing is called microgravity, not zero gravity, because they're actually falling. To help with this, think back to Newton's cannonball thought experiment. The cannonball was continuously falling to Earth but missing the planet as it curved away. The ISS is doing the same, and because they're inside the ISS, the astronauts are doing exactly the same. Everything is in free fall, but because the "room" the astronauts are in happens to be falling at the same rate they are, as far as they're concerned, they're weightless. If you're ever in a falling elevator, you can try it for yourself. While a lot of people will be screaming and freaking out, if you can keep your head, you'll be able to push up off the floor and "float," for a little while at least. You can also experience the same feeling on a roller coaster, or in a plane when turbulence causes a sudden drop in altitude.)

Back to gravity and weight. If you're used to Earth's gravity and end up on a planet with more mass, you'll feel heavier. You'll feel like you weigh more, because you would. Your arms would be heavier, along with your legs. Just moving would take more effort. Everything would be heavier, even the stuff you brought with you from Earth that you could always pick up and move around back home. Also, Earth-designed machines would use more energy to get the same jobs done as they do on Earth.

Inside your body, the changes to your overall health and physiological functionality would be worse. Your heart would work harder, since it would be fighting against more gravity than it's used to in order to keep blood moving from your feet and your head to your heart. Breathing would be more difficult, as your diaphragm would have to work harder to change the volume of your thoracic cavity to pull air into and push air out of your lungs. Your skin would sag more and sooner than it does on Earth, since collogen would have to fight against more gravity than it was designed for. You'd get wrinkles. You'd weigh more and get wrinkles! Double downside.

This is why the phrase "Earth-like planet" sends a thrill down the spines of exoplanet hunters. An "Earth-like" planet with the presence of an atmosphere, water, and roughly the same gravity could have life on it that we'd be able to understand: animals with similar evolutionary responses to gravity. Plants whose root systems are stimulated by gravity to grow down while the rest of the plant hunts for the sun.

That's not to say that life can't develop on planets or in environments with different gravity than what we have on Earth. After all, gravity hasn't been a problem for a lot of Earth creatures—specifically, the ones in our oceans. Thanks to the buoyancy of water, the effective "weight" of aquatic or marine animals isn't as much of an issue as it is for terrestrial animals when it comes to size. Our oceans are home to the largest animals on Earth for a reason: they don't need to fight gravity as much as us landlubbers do. But if you took a blue whale out of the ocean, it would die very quickly. The animal's breathing would quickly become impossible, blood would pool on the side closest to the ground, circulation would slow, and organs would be smashed by everything else on top of them. It would literally be crushed by its own weight.

FREE FALL: MORE THAN A PETTY THING

This is where Froopyland comes in. If you were creating an extra-dimensional playground for your young daughter, one of the first things you'd do (along with building bouncy ground and breathable water) would probably be to turn the gravity down to something a little less than that of Earth to lower the impact were your daughter or one of her friends to fall.

One of the cool things about gravity is that, since there's a constant number involved ($9.81 \text{ m}/\text{s}^2$), the math is pretty easy to play with. In physics, there's a set of well-known equations called the kinematic equations that can be used to mathematically model and describe motion. Playing with the properties of the following equations can help you model and calculate free-fall variables.

The two equations look like this:

$$v = gt$$
$$x = 1/2gt^2$$

* v is the downward velocity of the falling object, measured in meters per second.
* g is the acceleration due to gravity—again, $9.81 \text{ m}/\text{s}^2$
* t is time, measured in seconds
* x is displacement, or the difference between starting and stopping points; it can also be represented by y, since it's a vertical measurement—or, even more simply, it's height

From those two formulas, any aspect of the fall of an object can be calculated.

Want to know how fast something is falling after twelve seconds? Multiply twelve seconds by $9.81 \text{ m}/\text{s}^2$.

Want to know how tall something is if you know how long it took for the object to fall from it? Use the second formula and put your time value in for t.

Want to know how long it will take an object to fall fifty-eight meters? Rearrange the second formula to solve for time: $t = \sqrt{(2x/g)}$.

($t = 3.4$ seconds, by the way.)

Find yourself on an alien world and want to know what the value is for the acceleration due to gravity? Second formula, rearranged: $g = 2x/t^2$, and solve for g.

Try it yourself: time how long it takes Rick to fall off the cliff at the beginning of his and Beth's adventure in Froopyland, estimate the height of the cliff, and calculate the value for Froopyland's gravity.

For a bigger problem, calculate the downward velocity of Mrs. Pancakes, Rick, or Morty at any point during their fall after jumping from the airplane in "Lawnmower Dog."

The neat thing about gravity is that you're not just calculating the velocity of any object falling from a certain height, you're calculating the velocity of *every* object falling from that same height.

WHERE DID GRAVITY COME FROM ?

Gravity is one of the four fundamental forces, along with electromagnetism, the strong nuclear force, and the weak nuclear force (yes, those are the technical names). Despite the name of the last one, gravity is the weakest of the four forces. Just look at magnets on refrigerators. That tiny magnet is sticking to the refrigerator due to electromagnetic force. It has the gravity of a whole planet pulling down on it, and it shrugs it off like it's no big deal.

Gravity has been around since virtually the start of everything, splitting from other fundamental forces around 10^{-35} seconds after the Big Bang, and will be around until the end of everything, when all the matter in the universe is evenly spread apart and cannot be pulled back together into large clumps. It's responsible for the formation of the first stars, and still performs that task today, along with causing bits of matter to clump together into larger and larger chunks until comets, asteroids, and planets are formed. All are held in their respective orbits by a delicate balance of gravitational forces.

But what causes it? This is where we need some help from Einstein.

In his theory of general relativity, Einstein posited that all forms of energy (and mass is just "frozen energy," according to

$E = mc^2$) change the curvature of the space around them. So, while the Earth orbits the sun following the math laid out by Newton, the cause of this motion is the sun's changing the shape of the space-time around it. To think of it in a two-dimensional way, imagine a large, flat rubber sheet. Now put a weight in the middle of it, to represent the sun, and let it pull down that area. Next, roll some marbles around the sheet. Depending on where they started from, some will go off the sheet in various directions, but some (in this case, replicating the planets of the solar system) will move in a circular motion around the weight along the curve in the rubber sheet.

Your marble will ultimately roll down into the depression the weight caused, but that's due to decreasing velocity caused by friction between it and the rubber sheet. Otherwise, in the right circumstances, it would keep spinning around that depression in the sheet forever. Plucking the weight out of the depression and watching the sheet ripple as the waves pass through it also correlates with gravity.

If we go back for a minute to the Law of Universal Gravitation, that denominator, r, means that gravity exists between all things in the universe, since there is a distance between everything. Take that to the extreme, and it means the reach of gravity spans the entire universe.

Every fundamental force has a particle or field that mediates it, or transfers the force between things. Essentially, each force has a particle that is a carrier. For the electromagnetic force, it's the photon; the strong nuclear force has the gluon, the weak nuclear force has particles called W and Z bosons, and gravity has the graviton.

The exact specifics of the graviton in physics are still a little fuzzy right now, and no scientists have yet been able to make gravity work with quantum physics, though the other three fundamental forces all fit. This isn't to say that our view of gravity is broken; it's just that more work needs to be done. It's currently hypothesized that when evidence of gravitons is found, they will be massless and propagate at the speed of light, which would agree with indirect measurements of the speed of gravity. And the ripples in that rubber sheet would be the gravitational waves in the model.

Gravitational waves, which were originally predicted by Einstein in 1916, can be produced by catastrophic events between massive objects, such as a collision of black holes, or a single star collapsing in on itself. In either case, the gravitational effects of the event are enough to compress and stretch out space-time itself, thus making waves, albeit on the subatomic level. These waves are what was observed by the Laser Interferometer Gravitational-Wave Observatory (LIGO) in September 2015, when it registered a collision between two black holes nearly 1.3 billion light-years away. No need to worry, though: by the time the waves reach Earth, they're billions of times smaller than they were when they were originally created.

The observation of gravitational waves was the start of a new way of observing the universe, a new "sense" with which we can look out into the space around us. Gravitational waves are completely unlike electromagnetic waves, which include X-ray, radio, visible light, infrared, gamma, and other waves by which we observe the universe. The team responsible for the discovery in 2015, which verified Einstein's nearly one-hundred-year-old prediction, justly won the Nobel Prize for Physics in 2017.

What makes gravity so fascinating is that it only attracts and does not repel. Unlike electric charge (which has two sides, positive and negative) and magnets (which have north and south poles), there's no "opposite" to gravity. This means there is no such thing as "antigravity," which, in theory, would repel rather than attract.

Which brings us back, at last, to *Rick and Morty*.

FAKE GRAVITY

Every ship and space station in *Rick and Morty* has some form of artificial gravity. But none of them are spinning.

We already know how to simulate gravity in spaceships: just build something round and spin it or accelerate it at the right magnitude, and the acceleration will be indistinguishable from gravity. But that's not seen in *Rick and Morty*. As in most science fiction, artificial gravity is a given in *Rick and Morty*. No one must float or deal with the issues that come from zero G (or, as discussed earlier, microgravity) in space.

In *Rick and Morty*, artificial gravity is presumably provided by some material or plating that can be adjusted and tuned to provide just the right amount of gravity, pointed in just the right direction. Maybe someday, when we better understand gravitons (if our understanding of them is proven to be correct), we may be able to theorize ways to manipulate them in the same way.

All that said, replicating gravity in space isn't impossible for us. Accelerate your spacecraft at 9.8 m/s², and you invoke Einstein's equivalence principle. Inside the isolated system of your ship, that accelerated motion will perfectly mimic gravitation, and there's no way you'd be able to tell the difference experimentally. The constant acceleration would use a lot of fuel, though, and any changes to the acceleration—including, say, going a constant speed—would affect the gravity felt by the crew.

Another way to mimic gravity is to design a part of your spacecraft to be circular and set it spinning, a time-honored approach in more realistic science fiction. This is artificial gravity via centripetal force, which is directed toward the center of spin by any object that's moving in a circle, including the floor of your spacecraft's ring. If you're near the outside of that ring, the floor would push against you, mimicking gravity. Make the ring big enough and spin it at a speed appropriate for its size and you can create a force pointed at the center that will, again, imitate gravity.

Like the use of energy to "create" gravity via the equivalence principle and the other issues that would crop up when changing your acceleration, creating artificial gravity by spinning also has limiting issues. If the ring is too small, the "gravity" felt at a person's feet (on the ring) would be different from that felt at their head. Dizziness would be the least of that astronaut's problems. Additionally, you would need to keep the ring moving (with good propulsion) and make sure it's not moving so fast that it rips itself apart (with good engineering).

A GRAVITATIONAL CASE STUDY: DWARF TERRACE 9

In "The Wedding Squanchers," Rick and the Smiths settle on what he calls "Tiny Earth," also known as Dwarf Terrace 9. The name is pretty self-explanatory. And with what we now know about gravity, some interesting numbers can be calculated.

The morning after landing, when Rick comes in and asks if he's smelling bacon, the family tells him that they have discovered (and subsequently made extinct) a species of tiny pig off the coast of "New Australia," which, according to Beth, is "about thirty yards east," to which Morty adds, "Or three hundred yards west."

If Morty and Beth are describing the same point, and we know its distance from where they are, both east and west, then we know the distance around the full sphere, or the circumference. We're going to assume Morty and Beth were talking about thirty and three hundred yards due east and due west, respectively, which means the full circumference around Tiny Earth is 330 yards (302 meters).

Now we can calculate the gravitational field strength of Tiny Earth, using $g = Gm/r^2$.

We've got the circumference of Tiny Earth, and that's it. We'll need to calculate everything else for the formula above. To start off, we can work backward with the formula for circumference to calculate the radius.

The circumference of a sphere is the same as the circumference of a circle, so we'll use the formula:

circumference = $2\pi r$

Rearrange to solve for r:

r = circumference/2π
r = 302/ 2π
radius = 48 meters

Now we need mass. This part has a couple of steps.

We know the radius of Tiny Earth, and from that we can figure out the volume of Tiny Earth. Rick said that these three planets were all at least 90 percent like Earth. To keep things simple, let's assume that Tiny Earth is 100 percent like Earth. The Earth is a rocky planet, and therefore has a known density—and in this thought experiment, Tiny Earth would have that same density. And as we all learned early on in science, density = mass/volume. The density of Earth is 5520 kg/m^3.

So, what's the volume of Tiny Earth? We need the formula for the volume of a sphere:

V = 4/3 πr^3

Plug in 48 for r, cube it, and complete the calculation.

Volume of Tiny Earth = 463,240 m^3

Now for mass:

Density = mass/volume
mass = (density)(volume)
Mass of Tiny Earth = 2.6 x 10^9 kg

Now that we have that, the gravity of the situation really comes into focus. Use the formula from earlier, and you find that g = 7.5 x 10^{-5} m/s^2.

That means it has next to no gravity. Remember, the gravitational field strength on the moon is a relatively robust 1.65 m/s^2 in comparison.

If this was Earth, this would affect everything. An astronomical body this small wouldn't have enough gravity to be able to hold on to an atmosphere. After all, the moon, which is roughly the same distance from the sun as the Earth, doesn't have enough gravity to have one, and it has about 22,000 times the gravity of Tiny Earth.

Also, as a result of the low gravity, staying on the surface would be an issue. Just for reference, the Earth's escape velocity—the speed at which you need to move upward in order to escape the gravitational field of the planet—is 11,184 m/s. We can easily calculate the escape velocity for Tiny Earth, using the following formula:

$$v_e = \sqrt{2Gm/r}$$
$$v_e = \sqrt{2(6.673 \times 10^{-11})/48}$$
$$v_e = 0.08 \text{ m/s, or about 8 centimeters per second.}$$

Jumping up would be overkill. That's much slower than an easy jog, which is about 3 m/s. If you didn't stay still, just walking would propel you off Tiny Earth.

And the weak gravitational field would also affect how quickly things fall. Applying the free-fall formula from earlier under these circumstances, we can show just how wonky that tiny bit of gravity would be if you dropped something from 2 meters (a little less than Rick's height) on Tiny Earth.

Using the free-fall formula, $t = \sqrt{(2x/g)}$, and plugging in 2 meters for x, we find that it would take 73 seconds for the object to hit the ground. Just for comparison, on Earth, an object dropped from 2 meters lands about 0.64 seconds after you drop it.

The only way the gravitational field strength can change would be for the mass of Tiny Earth to be a huge value. Here's how we can use the math to get Tiny Earth's gravity up to normal Earth's value of 9.8 m/s².

$$g = Gm/r^2$$
$$9.8 = (6.673 \times 10^{-11})(m)/(48)^2$$

Rearrange and solve for mass: we find that, for the gravity of Tiny Earth to be the same as real Earth's, the mass would be:

$$m = 3.39 \times 10^{14} \text{ kg}$$

From that, we can go backward and figure out what Tiny Earth is made from, by using our density formula. If Tiny Earth has the same gravity as Earth, we know what its mass has to be; and its volume hasn't changed, so we can calculate the density:

$$D = m/v$$
$$D = 3.39 \times 10^{14}/463,247$$

Density of Tiny Earth (to provide normal Earth gravity) = 7.3×10^8 kilograms per cubic meter

That's basically the same density as a white dwarf star (a star near the end of its life), which has roughly the mass of the sun with the volume of the Earth.

When it comes down to it, Tiny Earth isn't very Earth-like at all. But that's not quite it: living on Tiny Earth would be its own horror show. Gravity decreases the farther you move from the center of mass, so your head would feel lighter than your feet because those two parts of your body would feel different gravitational fields. While it was a decent hideaway in the series, life on Tiny Earth would be anything but a good time, even if there was bacon.

CHAPTER 15

Time

* ★ ✦ ★ *

In interviews, *Rick and Morty* creators Justin Roiland and Dan Harmon have said that they have no desire to do any time-travel stories in the series whatsoever. In an interview with CBR TV that took place at San Diego Comic-Con, Harmon called time travel "a real shark-jumper" and "just a dangerous toy to pull out." And referencing the ever-present box on the shelf in the garage labeled "Time Travel Stuff," both Harmon and Roiland dismiss it, saying that it's just some stuff that Rick's been tinkering with that doesn't work, and its position on the shelf is both literal and a little meta for how they view time travel in the show.

So, the reason not to include time travel in the series is essentially an aesthetic one, but when you make a show heavily steeped in popular ideas of science, you cannot help but step into the murky world of time.

Also, as we've discussed earlier in the book, in the reality of *Rick and Morty*, wormholes are real, and are probably the underlying science behind Rick's Portal Gun. In another episode, Rick and Morty travel through the galaxy like it's no big thing, and travel near objects that have wildly different masses.

All three of these concepts involve some manipulation of time. While there may not be straight-up time travel in *Rick and Morty*, just from doing what they do, they would feel the effects of time in a big way. To figure out how, we need to do a deep dive into time.

TIME'S ORIGINS

It's too easy a joke (and too unfunny) to say that time has been going on since time immemorial, so let's not.

Instead, let's look at the actual, serious question within physics of when time "started."

The easy answer is that time started at the point of the first cause and effect. Without time, there's no before or after, and if nothing is happening, then there's no reason that time would exist. The Big Bang is something that happened, the event where we can clearly see the "before" and "after" that kicked off all the befores and afters since.

But that might not be correct. If, for example, we consider the possibility of multiple Big Bangs throughout a preexisting cosmos, as we discussed in our examination of multiverses, then "time" as a concept of before and after existed prior to the Big Bang that started our universe. Another view looks at the "rebounding" universe model, suggesting the infinite time line of inflation and deflation of universes, which likewise sets up "before" as "before our universe."

The point here is that, while there are some ideas, it's difficult to pin down just when time began. Given that we have no means of observing the universe or larger cosmos prior to the Big Bang, scientists have set the "start" of "time" at the Big Bang, about 13.8 billion years ago.

Just as physicists can't pin down an exact beginning point of time, trying to define time has proven to be equally difficult—almost like trying to hold sand in your fist. Actually, holding sand in your fist is much easier than defining time.

For something that seems so simple, time holds many, many

mysteries. Why does it flow? Why does it go in only one direction, and why are we pulled along for the ride? Why can't we "remember" the future? Why can't we move through time the way we move through space?

Simple concept. Mind-blowing questions.

Time has been known and measured since humans became humans, and keeping track of time has always been one of our most important activities. Accurately keeping track of the days and the phases of the moon meant that crops could be planted at the right time of year. The position of the sun would inform hunting parties when to return home, so that they wouldn't find themselves wandering in the dark woods and possibly falling prey to other animals. Time is programmed into all life—plant, animal, and even fungal—from multicellular organisms down to single cells, from the simplest diurnal rhythms to the most complex fertility cycles.

In terms of a first real definition of time, we must look again to Aristotle, who concluded that time is the measurement of change with respect to the before and after. Everything changes continuously, and time is the way to count, or measure, that change. It's a good idea, and still largely works today. But then, why not take something simple and make it really complicated, as philosophers are wont to do? Aristotle's straightforward explanation of time leads to the question: What happens if nothing changes? Does time stop or cease to exist? Or, if you could somehow stop things from changing, would time then stop? To Aristotle, the answer was yes: time doesn't continue if there is no change. Aristotle also concluded that the past was gone and no longer existed, the future did not yet exist, and the "now" was not a part of time, so therefore time itself might not exist.

There have been books written about the Aristotelian view of time and space. This is not one of them. But it's safe to say that the question "What is time?" and the nature of its very existence was largely the domain of philosophy for centuries. Science was a little too young to take a serious look at time until the last few centuries.

About two thousand years after Aristotle, Isaac Newton began looking into the subject, and, while he kept some of Aristotle's ideas, he largely saw time as something that flowed equally in all places, "without reference to anything external." In other words, the universe had an absolute clock that counted the seconds, and those seconds were equal for everyone in all places. Time, as humans measured it, was imperfect compared to this "true" time. This could be most plainly demonstrated by the need to correct the calendar for lost seconds and days, such as in the case of leap years (which allow us to correct for the fact that a year is actually 365.25 days). It was our fault, not the universe's.

Thus spoke Newton, Science God of the seventeenth century.

Given Newton's reputation, and the fact that he was just so damn right about so many other things, his word was the correct version of time for over two hundred years—until "A Rickle in Time" post-credits guest star Albert Einstein said it wasn't. If you feel like you're having déjà vu, it's okay. Einstein updated a lot of Newton's ideas and figured out how physics applies at the scales of gigantic masses and terrific speeds.

By the way, there is no evidence that Einstein's work on time was inspired by being beaten up by self-righteous "scrotal-looking creatures" traveling from the fourth dimension.

Einstein's interest in time started before he was Albert "General Relativity" Einstein. He famously worked as a clerk in a patent office prior to finding physics acclaim and the resultant university postings, and much of his time there was spent reviewing patents proposing methods and technology to allow people to synchronize time between different places. As timekeeping technology (clocks) got better through the centuries and as humans discovered faster ways of moving people from point A to point B, Newtonian ideas about time began to erode. As any train passenger in the late 1800s would realize, times were a) extremely important and b) sometimes not the same from place to place.

While the development of technology that was better at keeping time truly did change the world on virtually every level, we're going to focus on Einstein's understanding of time as explained in his theories of general and special relativity. As he worked, Einstein realized that the passage of time was not absolute but depended upon the location and the motion of the individual making the observation. When two observers in different locations or conditions compare the passage of time relative to each other, it's not the same. According to Einstein, the passage of time is in the eye of the beholder. Nothing occurs simultaneously in the universe. There is no absolute "universal clock" that gives a time upon which we can all agree. This isn't like time zones. This is like time appearing to move faster in some areas of the universe and slower in others, relative to different frames of reference.

Einstein's work also showed that time is a crucial piece of reality, a fourth dimension added to the three dimensions of space that we experience. This came from his theory of special relativity, in which he showed that the laws of physics are the same for all nonaccelerating observers (as long as you're not accelerating, the laws of physics work like they should; but once you start accelerating, all bets are off), and that the speed of light in a vacuum is constant, no matter what speed the observer is traveling. Taken together, these two findings led Einstein to the understanding that space and time were not independent of each other but rather completely interwoven into something he called space-time.

However, a question from Einstein's day persists: Why can we move in any direction in space but in only one direction in time?

This is the so-called arrow of time, or, speaking in physics language, the asymmetry of time. Think of your coordinates in space. You are at the center of three axes: up/down, left/right, and forward/backward. In theory, you could move as far in any direction (and in any combination of directions) as you wanted. Or you could not move at all and be fixed in space, your location static.

Now think about time. You exist "now," and that time is a point on an arrow that points in only one direction: into the future. Everyone you know, and the entire universe as we understand it, moves in that same direction. There's absolutely no stopping, no backing up, and no hopping around, and the idea of different dimensions of time that would be related to the three dimensions of space is pretty mind-melting. And the whys and hows of the seemingly locked-in characteristics of time are the questions that modern physics is struggling to answer.

TIME'S ESSENCE

With Einstein's embedding of the inexorable nature of time within the larger universe, the concept (and term) of "space" has been slowly replaced by the more accurate "space-time." But still, that leaves the question "What is time?" hanging. Our current definitions are little better than Aristotle's (that time is a measure of change). And while the work of Newton, Einstein, and countless other physicists and philosophers has laid conditions and boundaries on what time can do, the idea of what time actually is—and, perhaps more weirdly, if it even exists—stumps physicists to this day.

Perhaps this is why time travel doesn't show up in *Rick and Morty*. Time could be the one thing that, even though he can manipulate it, Rick still doesn't quite understand, and thinks it's best to just leave alone.

From our perspective, in both our individual and collective experience of it, time could be seen as the thing that stitches together all the causally connected but static moments of experience. The cause-and-effect nature of the universe is perhaps the strongest evidence for the existence of time. In our universe, cause happens before effect. If it didn't, the universe just wouldn't make sense. To give a very simple example: a ball is thrown at a window, and the window breaks. The ball being thrown caused the window to break. Then, more broadly, we are living in the present, which is the effect of all the causes in the past. We have never seen—at least as far as we understand—an effect happen before a cause. All the laws of physics depend upon cause and effect. Finding evidence of the uncoupling of the two would mean a fundamental change in everything.

Another "proof" of time—or at least of the direction of time's flow—is related to entropy, or disorder, and how it increases per the Second Law of Thermodynamics. This is kind of like the universal version of cause and effect: ultimately, all the effects in the universe have more entropy than the causes. Wood burns, and the resultant parts (ash, smoke, heat) have more entropy than than the original piece of wood. Again, like cause and effect, we have never observed reverse entropy—that is, ash, heat, and smoke spontaneously reassembling themselves into a piece of wood.

In this sense, entropy guides the flow of time, leading the universe to more disorder and, ultimately, to cold, cold death (if that is, in fact, the ultimate end of the universe: one where disorder reigns supreme). The big question, though, is: Does entropy guide and nudge time in an exclusively one-way direction, or is it along for the ride while time does its thing?

Einstein's theory of general relativity throws a monkey wrench into our attempts to define time. His theory explains that time, as we best understand and experience it in the large-scale structure of the cosmos, doesn't work with quantum physics (the submicroscopic side). And the incompatibility gets worse. In exploring this incongruity and attempting to make the relativistic and quantum realms of physics work in harmony, physicists John

Wheeler and Bryce DeWitt developed an equation that was, and still is, controversial. In it, time disappears. Like, gone. At the very smallest, most fundamental level of the universe, time doesn't seem to exist. All the laws that govern the quantum realm just don't need time to work. At this extremely small level, there's no past, no future, and no "now." It turns out that the idea of "now" seems to matter only to humans, rather than to the universe as a whole. This weirdness is so well-known and established that it has its own name in physics: the problem of time.

This discovery and the resultant research on time in the decades since could be as significant as Copernicus's realizing that the Earth revolves around the sun rather than the other way around. Everything we thought we knew about time (which, admittedly, wasn't that much) might be wrong.

The problem of time has led some physicists to conclude that, yes, time isn't real as far as the universe is concerned, but it's something that's created by observers, so our notion of time is neurologically created. This view, its proponents point out, gives a reason for Einstein's claim that time is relative to the observer; the observer creates time, and all devices that we use to "tell" time further reinforce the creation of time by us, the observers.

While this view may seem odd, to say the least, quantum physicist Spiros Michalakis points to personal evidence that supports the idea. "If that link to what tells us that something has changed is severed, we cease to exist," Michalakis says, "even though we're still part of the universe relative to other people for whom that link is still active. From our point of view, time, space, and all of existence goes away. This is something I've experienced twice when I've gone under general anesthesia.

"When this happens, you are removed from the link that tells you things can change. That stops your ability to register change, and when that happens to you, it's not like you go into a dream or you're just asleep, where you can still register change; 'you' just cease to exist. Maybe three and a half hours pass, but when you wake up, you have no idea that any time has passed."

As Michalakis sees it, the link to the knowledge that something has changed is in the brain. Sever that, and you, in a sense, cease to exist.

Whoa.

Ultimately, clear understandings of time and its flow may involve bits of relativity, quantum theory, neuroscience, thermodynamics, and more.

OG TIME LORD: ALBERT EINSTEIN

Let's go back to Einstein and his relativistic view of time. He showed that time moves at different rates for two people, relative to each other, if they are moving or are in different gravitational fields. This effect is called time dilation.

Let's look at motion (which comes as a result of special relativity) first.

Your twin leaves for space for a year, traveling at 99.999999 percent the speed of light. Six months out, six months back. Your twin would have aged one year, as time passes normally for them on their ship. But when they open the door of their ship after coming home, they will have found that roughly seven thousand years have passed, and you will most likely no longer be alive.

This happens, as Einstein realized, because the speed of light is the hard-and-fast constant in the universe. It's an unshakable 3.0×10^8 m/s everywhere. Nothing can go faster than light, "warp speed," "hyperspace," and every variation in between be damned—there are no cheats and no workarounds. Let's also go back to thinking about moving through time as a fourth dimension, and that speed limit is also the speed of light.

First, think about you at home, on Earth, having just said goodbye to your twin (for the purposes of this thought experiment, you have a twin, and you've just said good-bye). Relative to the Earth, you're not moving through space, so as a result, your speed through time is large, in order to reach that set speed limit. Now for your twin: they're moving with a lot of speed through space, but they've only got a set amount of speed, so their movement through time is small (at least, as measured by your clock on Earth). On your twin's spaceship, time is passing normally. Who's right about the passage of time? Both you and your twin are, although you'll miss each other by several millennia when they come back home and want to discuss it.

Two observers can experience the flow of time differently, and this has been shown experimentally numerous times since Einstein revealed it. Superaccurate atomic clocks have been synchronized and then compared after one was flown in an airplane and the other remained on the ground. The airplane clock reported that slightly less time (on the order of milliseconds) had passed than the clock on the ground did. In human terms, astronauts who spend a large amount of time on the ISS are technically "younger" (again, on the order of milliseconds) than they would have been had they stayed on Earth.

Calculating the difference in times is easy, given the proper variables, using the formula $\Delta t = \Delta t_o / \sqrt{1-(v/c)^2}$.

* Δt is the observer time, the time elapsed for the stationary, or Earth-bound, observer, in seconds.
* Δt_o is the "proper" time, the time elapsed for the fast-moving object or person, in seconds
* v is the velocity of the moving object, in meters per second
* c is the speed of light, 3.0×10^8 m/s

Since the speed of light is involved, special relativity rules are invoked, and the speeds are called "relativistic" speeds, meaning that they are a significant fraction of the speed of light. Input how much time the fast-moving observer reports between events (the proper time)—for example, starting and stopping a stopwatch or beginning and ending a journey—along with the values for their velocity and c, and you're able to solve for how much time passed for the Earth-bound observer. Subtract the observer time from the proper time to see how much more or less time passed for each observer.

This can be done for any velocity, but obviously, and just from looking at the formula, the smaller the velocity of the moving object, the smaller the difference between the two. For ISS astronauts again, the difference is in milliseconds, and the station moves at 8 km/second. You're not getting younger on your daily commute—unless you go really fast.

Speaking of going really fast, let's bring Rick and Morty back into the picture. Space is big, and traveling through it takes time. A lot of time. To cut down on that time, you can do one of two things: invent portal technology, or invent the technology to fly your spacecraft at relativistic speeds. Rick has done both.

Given the average distance between planets and solar systems in our galaxy, you'd need some means of traveling by shortcut, or the ability to go really, really fast if you wanted to get anywhere in the space of, say, twenty-two minutes (without commercials). To be fair, Rick and the Smiths most likely use a combination of portal and sub-light-speed travel to get to their destinations. Rick's Portal Gun does plug into the ship's dashboard, after all. But still, travel between planets or between solar systems without portal tech? You would need relativistic speeds, and it's the people not traveling at relativistic speeds who are the ones paying the piper for your super-high speeds.

So, if Rick is traveling at relativistic speeds, he's performing at least a type of time travel, and one that someday will be available to us. If you need to jump into the future, use the formula above, fire up your ship, zoom out at, say, 90 percent the speed of light for a set time, turn around and come back, and boom—you're in the future of the Earth you left. You will have aged by however long it took you to travel, of course, but time on Earth will have moved faster from your ship-bound frame of reference. You'll be in the future.

Einstein's other method of permissible time travel came about as a consequence of special and general relativity and has everything to do with mass. As you move closer to a large mass, time passes slower for you, relative to someone not in your specific frame of reference—say, back on Earth. In more science-type terms, clocks in a strong gravitational field move slower than clocks in a weak gravitational field. Gravitational time dilation appears in accelerated frames of reference, and even though the surface of the Earth may not be accelerating in the sense of motion, the equivalence principle states that an accelerated frame of reference is indistinguishable from a gravitational field. For a refresher, check out Fake Gravity in the last chapter.

Like time dilation based on motion, this version of altered time has been experimentally shown numerous times in many different conditions, starting in 1959 when researchers took one of a pair of superaccurate clocks to the top of a mountain and let it run for a set period. When it was brought back down to the surface and compared to its twin, more time had passed for the clock that went mountain climbing. Given the overall age of our planet, the core of the Earth (with a stronger gravitational field) is 2.5 years younger than the surface. By the time you die, your feet will be a few milliseconds younger than your head.

While we've never seen Rick encounter a truly massive object in the universe, like a black hole, he's mentioned them, so we know they exist. Gravitational time travel, like that based on velocity, would be one-way, and it would appear as time travel only in the eyes of the observer. You'd just have to fire up your ship, head out into the galaxy, and park it a few kilometers away from the event horizon of a black hole that's maybe a thousand times as massive as the sun (just make sure you don't fall in). To people on Earth, you would appear to move slower and slower until you froze, but on the ship, you'd be living your normal life, if not a little more nervously given your proximity to a black hole. Stay parked for around a year and then head home. Open your ship's door and everything you would have remotely recognized on Earth would be gone.

Big science-fiction picture: think of life on planets that orbit massive stars or black holes, or on planets that move through space much more rapidly than the Earth moves around the sun. Time for them is not time for us. Remember that time is not an absolute. There is no "universal" time, where one Earth second equals one Gazorpazorpian second. There's no universal "now" in the universe, and therefore there's no common history for the universe. Time is relative.

In our world, time dilation hits all of us right in the pocket. No, not your wallet—the other pocket. Your phone.

The GPS signals our phones receive and that we depend upon to tell us where we are come from satellites overhead. Just like the astronauts on the ISS, Global Positioning System satellites are subject to the effects of time dilation: their clocks run slightly slower than surface clocks because of their orbital velocity because of the lesser gravitational field strength. As a result, the clocks on the satellites must be constantly adjusted to reflect time on the Earth's surface—otherwise, the differences in time would pile up, ultimately making GPS useless.

Now might be the right time to discuss breaking time.

STOP. SANCHEZ TIME!

It's time to talk about Rick "freezing" time in "Ricksy Business" and "A Rickle in Time."

Put simply, it's not possible. In fact, it's a mess even to think about hypothetically. For instance, if you stop time, the atmosphere would stop moving as well. Moving, breathing, hearing, and staying warm would all suddenly be very important things. All those things depend upon the air in the atmosphere moving, and if time stopped, you'd realize that there would be no changes in the position of the air molecules around you. Aristotle would be proud that you remembered his definition of time, but you'd be dead.

If you got smarter and kept time moving in the atmosphere, you'd probably want to keep time running for the electromagnetic spectrum as well. That's the infrared energy that can move through space and warm you up, and the photons that make up the visible spectrum so you can see. Don't worry about radio waves to stream a movie to watch while the rest of the world has stopped, though, unless you let time move for the computers storing those movies in the cloud.

While we're talking about things that move through space, by stopping time in toto, you've really screwed things up if you like staying on the surface of the planet. By stopping time, you've just shut off gravity. As explained in the last chapter, gravity has a speed. For something to have a speed, it must have an initial position and a final position that changes over time. Take time away and nothing's moving from one place to another, not even hypothetical gravitons that carry gravitational force.

Stopping time, in short, is bad news. And this isn't even bringing in the ideas of Michalakis and others that suggest that if we become unmoored from time, we would cease to exist.

But as time started to fracture in "A Rickle in Time," Rick, Morty, and Summer were bumped outside their own time line and became entirely hypothetical. According to Rick, the garage was in a timeless oblivion that made it sensitive to the uncertainty expressed by Morty and Summer. As such, every uncertainty felt by any of the three time-freezers ended up splitting reality (as predicted by the many worlds hypothesis, which was mentioned in the Multiverse chapter).

Some respect is due here, though. Taking a leap into some totally hypothetical physics, blending freezing time with many worlds, a shout-out to Erwin Schrödinger and his cats, and a dude from the fourth dimension who can see and hear and experience everything going on in all the multiple realities is just all kinds of physics loveliness, so I salute *Rick and Morty*'s creators.

But you still can't just stop time.

TRAVELING THROUGH TIME

While we're talking about time, we should cover all the bases. Straight-up time travel is the name of the game.

We talked about how you could, hypothetically, travel into the future—although it's a one-way trip—with time dilation, as allowed by Einstein's general and special relativity. Given the uncertainty and unknowns that surround time throughout this chapter, at least according to Einstein's theories, more options for time travel should be possible, even beyond the dilation explained earlier.

And what makes this a little troubling is that the means by which time travel could hypothetically become possible are well-known to and constantly used by Rick Sanchez: wormholes.

Here's the theory behind traversable wormholes: open one up on your end and then open the other end near a black hole, or accelerate the other end to near light speed. Let them sit for a while as time passes normally for you on Earth. Then, when you've calculated that enough time has passed, step through either wormhole—the one near the black hole or the one whose end is moving at relativistic speeds. Now you're not just going from point to point in space; you're going from point to point in time. Relative to the people in the "wormhole station" you just left, time is now moving differently for you.

While this approach sounds great in theory, you may recall that we have yet to confirm that wormholes exist, and that even if they do, it seems pretty unlikely that traversable ones could be found or constructed. And that's only the most obvious reason why wormhole-based time travel has problems. A full understanding of traveling through time via wormholes would have to include quantum effects, and when such a hypothetical system was modeled with an approximation of quantum effects (in addition to the relativistic character of the wormholes), the conclusions in regard to wormholes' ability to allow objects to travel through time were inconclusive. Some results indicated that such a means of time travel would be possible, while others suggested that the time-spanning wormhole would destroy itself. A better understanding of quantum physics and wormholes is needed to make this work.

If we acknowledge, like the Galactic Federation has, that Rick is the master of portal technology, and that portals are, in fact, traversable wormholes, then that might explain why Rick has a box marked "Time Travel Stuff" in the garage. He's known how to travel through time all along. Maybe he's been time traveling between episodes, maybe something horrible happened to him, or maybe he caused something horrible to happen and just doesn't do it anymore. Or maybe he realizes that at least half of time travel trips would be, by and large, pointless due to cause and effect.

Going back to the idea of cause and effect and how, in our universe, all effects must have causes, let's look at the harmfulness of a more typical science-fictional trip to the past. If you went back to the past as a tourist, just to look around, take some pictures, and then come home, that would be harmless, as long as you don't take anything, don't interact with anyone, and aren't seen in public. Easy.

This approach to time travel—past tourism—was actually one of the reasons Stephen Hawking believed that functional time travel was impossible, no matter what the theories said. After all, Hawking postulated, if time travel does become an option at some point in the future, we, and every other time period before time travel's discovery, would be inundated with "time tourists."

The idea of time tourists was something that entertained Hawking so much that in 2009, the famed physicist held the first party for time travelers where invitations weren't sent until after the party. If time travel did exist, Hawking hypothesized, it wouldn't be any trouble to get an invitatation after the party and then travel back in time to the party itself. Alas, no one came to the party. No one, it seemed, had the means to get to a party that had already happened.

Also, when it comes to time travel, we've all seen the movies and read the books, and no time traveler is careful enough not to change anything. Something is always is altered by the traveler by their mere presence.

And here's the thing: look at this plot through the lens of cause and effect. If you're living in the present and you decide to travel back in time, the past has already happened. That's how time works; the past happened, and the present is happening now. If you go back and change something preemptively in order to change your present, well, your present as it is (without the change in the past that you're trying to make) would inform you that you didn't or can't change the past. The knowledge of the present you would take with you into the past is an effect created by the causes of the past. You can't put that effect in front of the causes. If that worked and time travel was a possibility, then we'd live in a world where people would pop randomly in and out of reality as changes to causes in the past resulted in new effects in the present. Every effect in the past led to another cause and on and on, and all of those together, across the globe, built our present world. Change one thing in the past and it will ring through to the present.

Likewise, consider the time-loop trope where time is stuck on repeat. In this trip, you have to fix something, but you must figure out what needs fixing and repair it before time starts moving forward again. It suggests that, somehow, you'll be holding on to information on your trips through time and applying that information to events in the past in order to fix the future and get time moving linearly again. You're bringing an effect back to change a cause.

If you travel in the other direction—into the future—and start messing with things, you'd have the same problem, but in reverse. Let's say that, by using your wormholes, you travel into the future, where a kindly old inventor gives you a portable, kitchen-top home fusion generator that provides enough electricity for your entire neighborhood from coffee grounds, apple cores, and banana peels. You bring it home to the present and use it, providing your neighborhood with electricity. Back up, McFly: it's not even clear that you could bring it home if you wanted to. After all, if you did, you'd be changing the past of the inventor and putting an effect in front of the cause. That future happened—you went there and it was real—so if you use the fusion generator, that future is different and not the one you visited.

You know, maybe Roiland and Harmon are onto something here. Rick may be smart enough to understand time travel, and possibly to lay the groundwork for a possible time machine, but it would just make for really messy stories.

NOVIKOV AND HAWKING: TIME COPS

Heavy-duty research and ponderings on time travel have led other physicists to a point of wondering what's "out there" that would stop or prevent a change in the cause-and-effect nature of our universe.

Two of the more intriguing ideas come from physicists Igor Novikov and Stephen Hawking, and there are many others.

While considering the strict causality that our universe demonstrates, Novikov hypothesized in the mid-1980s that the effects of the past were locked in and could not be changed. He proposed the self-consistency conjecture, which states that time paradoxes—changes in the present caused by alterations of the past—are impossible and could not happen. As such, Novikov said that the probability of any event that could cause such a paradox or a closed time line curve, as it's called in general relativity, is zero.

This result is one often seen in science fiction, and a lot of stories follow the idea that you can't change the past. To take a popular example: if you were to go back in time and attempt kill Hitler as a baby, it wouldn't happen. You would be delayed in your attempt. You'd be caught, and the opportunity would be lost. A car would hit you on the way to your murder appointment. Your hands would be injured, gangrene would set in, and both would be amputated. Any co-conspirators that you recruited in the past would find their attempts likewise stymied. Ultimately, even if you did kill Hitler as a baby, it wouldn't be baby Hitler that you killed, and Hitler would, somehow, still rise to power, causing all the effects that played roles in creating your present.

While the example is both common and extreme, if Novikov's larger ideas about the immutability of the effect side of things is true, then free will goes away, as we're all just walking piles of effects from all the causes before us.

For a broader approach to the problem, Hawking, who loved the thought of time travel, gave a name to his ideas: the Chronology Protection Conjecture. While hedging his bets and acknowledging that time travel may, in fact, be theoretically possible, Hawking limited the travel to submicroscopic particles, claiming that the laws of physics would prevent anything larger from making such a journey, and thus that any time machine that could be built would not work. Playfully, Hawking wrote, "It seems there is a Chronology Protection Agency that prevents the appearance of closed time line curves and so makes the universe safe for historians."

Critics of Hawking and Novikov's view of chronal protection of one form or another, as well as their overreliance on time travel prohibitions, point to the fact that all of the above is applicable only if there is a single, nonbranching time line that must be preserved. In the opposing view, which brings in ideas from the many worlds theory, the inflexion points that branch off to new realities aren't limited to the present moment. If many worlds is a manner by which alternate realities are created, then the whole time line should be open to the possibility. Go back in time and prevent Kennedy from being assassinated, and now there are two time lines, each forming their own reality: one in which Kennedy did not die in Dallas, and the original one that you came from, where Kennedy did die. The trouble comes when you return "home" to the present day. Could you stay in the time line where Kennedy didn't die, even though there might be another version of you in it? Or are you limited to travel in your original time line only, and the changes you made are experienced by an entirely different reality?

Maybe that's the reason there's no time travel in the show: each trip would create new realities, and with an infinite number out there, there's more than enough already.

CHAPTER 16

Growing and Shrinking People

★ ★ ★ ★ ★

Like any good science fiction story, *Rick and Morty* shrinks and grows people with some regularity. So far, each of the first three seasons of the series have had episodes in which characters such as Morty, Ethan, and the president have shrunk to microscopic size or grown to truly epic proportions, like Summer, Beth, and the dear departed Reuben.

And this is all done by Rick, who has the technology to make people small and large, and, as he snarkily points out to the president in "The Rickchurian Mortydate," his tech works without endangering the health of the user, while the president's shrinking pills come with a shockingly high chance of getting cancer baked in.

In our world, though, size isn't something that can be changed on a whim. It's part of any species's physiology and anatomy. Change it and you invite in all kinds of problems.

We're going to start with the small.

MAKING A SERIOUSLY TINY RICK

In both "Anatomy Park" and "The Rickchurian Mortydate," characters were shrunk from normal human size to something much,

much smaller. The president is shown shrinking (right out of his clothes, which don't shrink along with him) to maybe a quarter of an inch, while Morty shrinks far smaller, to be able to fit inside Reuben's body and enjoy the wonders (and horrors) of Anatomy Park.

There aren't actually any methods for shrinking people in reality, so let's consider why there aren't.

As best we can logically understand it, there are two ways to shrink within the bounds of reality: 1) smoosh everything together into a smaller volume, and 2) remove a portion of the object's mass.

Smooshing all the matter in a body together into a smaller package would mean that the atoms making up the person would have to be made smaller. Just smooshing the body down would damage organs and tissues and likely kill the individual. And, on the scale of the very small, pushing the atoms closer together is not an option, since like charges repel. Electrons, which are in a "cloud" around the nuclei of atoms, don't like to be near other electrons, and if a force could somehow push on neighboring atoms to make them snuggle up closer together, those electrons would push back just as hard. Finally, a case against smooshing is made with one word: density. If you decrease the volume an object occupies but keep its mass roughly the same, the object's density will increase dramatically. While an increase in density would cause multiple biological problems, it would also have the effect of causing the smaller object to push into and through any surface upon which it rested. Shrink an object small enough while keeping its mass the same and you can easily end up with a small package with a density such that it would sink through the Earth, stopping only when it reached the center.

To make the functioning person smaller, we'd want to make everything smaller in correct proportion to everything else. That means the atoms themselves would have to shrink. And that's also a problem.

Atoms are made up of a nucleus, which contains the positively charged protons and neutrons (which carry no charge), and a cloud of electrons, which zoom around the nucleus in probabilistic locations within orbitals at extremely fast speeds. Because the electrons are in constant motion around the nucleus, the bulk of the atom's volume at any given time is just empty space. But that's not an opening to make atoms smaller.

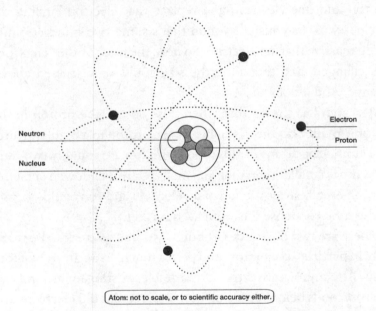

Atom: not to scale, or to scientific accuracy either.

As best we understand the physics of atoms in that subatomic, or quantum, realm, the distance between the electrons and the nucleus, as well as the distance between the protons and the neutrons inside the nucleus, cannot be changed when it comes to regular matter. The size of the atom is based upon several factors, such as the number of protons in the nucleus, the charge and mass of the electron, a universal constant called Planck's constant, dielectric permittivity (which controls the strength of an electric field), and 4π. Electron mass and charge and Planck's constant are locked in and cannot change. The atom's size is fixed.

Even inside the atom, there are rules that must be followed. Although, since Rick can shrink people, as he does in "Anatomy Park," "tuning" the constant in a specific field might be the way he would go.

Although research seems to suggest that, under the right conditions, the rules that govern the size of atoms—hypothetically—can be bent. Take a normal hydrogen atom, made up of one proton and one electron, and replace that electron circling the nucleus with its cousin, a muon that's some two hundred times more massive than an electron. So now, the mass of the "electron" has changed, and therefore, the size could, as discussed above, change. And it does.

Research has shown that when this is done, the proton in the muonic matter becomes smaller than a proton in a regular hydrogen atom. The atoms of this muonic matter are roughly two hundred times smaller than their normal, electron-endowed versions. Find an even heavier version of an electron and you could—again, in theory—condense the atom even further.

There are two downsides, though. Muonic matter can be made only in particle accelerators and pretty much on an atom-by-atom basis. Also, muons are extremely unstable, existing for around two microseconds before decaying back into regular old electrons and a couple of other subatomic bits.

Making the atoms of a person smaller also introduces an inherent problem. Just making the parts of the body smaller does nothing to their mass; it affects only the space they occupy—their volume. Density is the ratio of mass to volume, and if the mass of an object remains the same while the volume decreases, the density has to increase.

In other words, shrink a normal human to action-figure size and now that (on average) seventy-five kilograms of mass is in a much smaller space—and you have a much-denser-than-normal tiny human. Quick calculations for the densities of shrunk humans put them in ranges that are more often seen with cosmological objects such as neutron stars. With that density, the shrunk person would break through the floor, and continue breaking through whatever was beneath them until they reached the center of the Earth.

So, making the atoms in a body smaller isn't going to work. Removing some parts of them in order to make the individual smaller doesn't sound like a winning plan either, when you have to consider which atoms to remove. It's not as if bodies have extra atoms that aren't part of anything important just lying around, waiting to blink into and out of reality as needed so the owner can get smaller.

In order to reduce the size of a person by 10 percent you would have to take away 10 percent of their atoms. Again, which ones? The ones that are in molecules of DNA, which would leave the molecules useless and potentially harmful to the cell they're in? Atoms that are part of neurotransmitters in the brain? Atoms that make up the molecules in the components of muscle fibers? Even if this were a possibility, the location of the missing atoms is an unanswered question, as is getting them back in the right spot.

Shrinking a person is still, and will probably remain, science fiction. And that's a good thing, because being small would seriously suck.

SMALLER IS NOT BETTER

The big thing to remember in all of this scaling (and we'll talk about it in the growing-bigger section as well) is that the surface area changes by the square, and volume (and therefore mass) changes by the cube. This is called the square-cube law.

Take, for example, a six-foot-tall person. Shrink them by half. Now they're three feet tall, or they've been shrunk by a factor of two in their linear dimensions. The person's surface area has decreased by the square of our change factor, 2 x 2 or four times, while

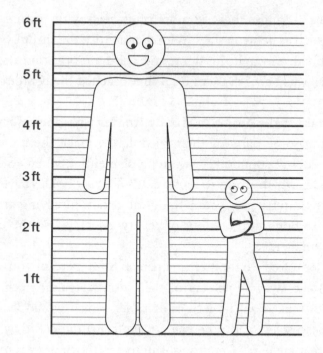

the person's volume has decreased by the cube of the factor, 2 x 2 x 2, or eight times. You can also run this problem by using 0.5 as the shrinking factor (6 x 0.5 = 3), so therefore their surface area changes by the square 0.5 x 0.5 = 0.25 of the original, while their volume changes by the cube 0.5 x 0.5 x 0.5 = 0.125 of the original.

Despite having their size cut only in half, the now-smaller person has only one-quarter of the surface area they once did, and only one-eighth of the volume and mass they had originally. That's going to cause some problems for them.

Think of a small insect—an ant, maybe. Ants are not set up physiologically, anatomically, or in any other way like human beings, and there's a reason for that. Functionally, ants live in a different world than we do—one where gravity isn't as big a deal as it is here, and where just breathing and staying warm would be extremely difficult, if not impossible, for a shrunken human being.

Let's say we took our six-foot person and shrank them to one inch tall. That's a reduction in size by a factor of seventy-two, so the surface area of the body would be down by a factor of 72 x 72 or 5,184 times, and the volume of the body would decrease by 72 x 72 x 72 or 373, 248 times.

Let's start inside the body. Vital bodily functions such as breathing, absorbing nutrients, and heat regulation are all proportional to the surface area of the organ doing the work, while consumption of oxygen and nutrients, as well as the production of heat, are proportional to the volume. As a result, one of the first problems would be maintaining body temperature. Even though the person's surface area, through which heat is lost, decreased only by a little over five thousand times, heat is produced in proportion to the volume the person, and there's over 370,000 times less of it than there used to be. Maintaining body heat would be a critical issue. To beat that, the person would need to take in more food to produce enough heat. But there's a complication there as well: the surface area of the gut and the lungs have both been reduced by a factor of 5,184, so they're not bringing in as much nourishment and oxygen, respectively, as they once were. Breathing faster should help maintain adequate blood oxygen, but eating would need to be a 24/7 thing just to keep the energy up. But if the person is active—say, getting into fights while trying to escape from a giant human body—they would be generating more heat, with less surface area to get rid of it; they'd be burning up. The person would also be losing water at a proportionally larger rate as well.

Vision would be difficult, to say the least, as smaller eyes mean smaller pupils, which won't let in as much light; and also, the wavelengths of light that do make it through the much smaller pupil would diffract as they come in, resulting in an unfocused mess. The shrunken individual's sight would be blurry and dark and would only get worse if they keep shrinking.

Just like the president's (and many characters in science fiction before him), their speaking voice would be shrill, as if they'd just sucked in helium, because their vocal cords would be smaller and therefore vibrate faster than they did when they were their normal size. Think of the vocal cords as vibrating strings: decrease the length of the strings and they vibrate faster. The sound of the small person's voice really doesn't matter, because regular-sized people would never be able to hear it. The volume at which you produce sound (the amplitude of the sound wave) is produced as a function of the pressurized air you push across your vocal cords. Smaller lungs, less air, less amplitude, less sound volume.

And, finally, our shrunken person wouldn't be able to hear our voices. The small cilia inside our ears, which vibrate as a result of sound energy, would be much smaller in the one-inch-tall individual, and, as with the shrunken vocal cords above, would vibrate at a faster rate and have a cutoff at around 300 hertz. As a result, the shrunken person would not be able to hear the range of normal human speech, which is around 200 hertz.

On the plus side, though, the smaller person would be proportionately stronger than they were at full size. This is due to the fact that strength (that is, forces produced by muscles) is proportionate to the cross-sectional area, which has only decreased by the square, unlike the volume (the body mass that you need to move around with the muscles), which has decreased by the cube. In other words, the person would have extra muscle power, because moving their body's mass isn't as big a deal as it was when they were full-sized. Lifting objects fifty to seventy times their (smaller) body weight wouldn't be difficult at all. Ants can do this easily, as their muscles have a greater cross-sectional area relative to their body size when compared to humans or larger animals.

Another plus would be the risk of falling. The gravity on falling small bodies increases less rapidly than the drag of the air through which they are moving, and as a result, the velocity at which they fall decreases. Also, less mass means less energy upon impact, so you could probably fall from a great height and not suffer too much injury.

The weirdest thing that the small person would encounter, though, would be water. Because of the person's greatly reduced mass at this scale, the forces that create surface tension are stronger than gravity. This would make water dangerous and drinking difficult. Liquid water would exist in blobs that larger individuals would recognize as dew drops. The surface tension would overcome gravity to maintain the blobbish shape, if the person tried to scoop up a handful of water. As for drinking, the water in the blob would want to stay "connected." Trying to sip the blob would feel like drinking something with the viscosity of mucus. There would be no small sip—just a start, and the rest would push its way down your throat, whether you wanted to drink it all or not. Do it wrong, and just drinking water while small could load up your stomach and fill your esophagus, keeping your epiglottis closed, resulting in your suffocating to death. Long story short (or maybe tall story short?): being shrunk would not be a good time. But don't be fooled—growing doesn't make things better either.

THE MOTHER-DAUGHTER GIANT TEAM

In "The Whirly Dirly Conspiracy," playing with size goes the other way, as Summer accidentally turns herself into a towering giant, and, to help her, Beth does the same.

As with shrinking, there's just no way to make this work technologically. That part of this will most likely always be fictional. But if Summer and Beth did manage to somehow enlarge themselves, being a giant would still be no treat, and being turned into a giant from normal human proportions would be as deadly as shrinking. Again, it's all due to the square-cube law.

Let's say that as a result of her growing, Summer's final height was one hundred feet tall, up from approximately five feet, five inches. Divide her final height (100 feet) by her regular height (5.42 feet), and we find that giant Summer is 18.5 times as tall as normal Summer. If we also approximate Summer's weight at 120 pounds, we can figure out some details about giant Summer.

Summer's surface area would see an increase by the square of 18.5, or 342.25 times, while her volume would be increased by the cube of 18.5, or 6,332 times. If we're consistent with our explorations of a shrinking person and claim that Summer's mass would, somehow, remain the same despite the massive change in her volume, then we've got a problem: that much mass in a gigantic volume means that giant Summer's density is next to nothing. Literally. She would have a density similar to that of a cloud. And, as such, her strength would be next to nothing as well.

If her mass increased along with her volume, and her weight was 120 pounds at her normal height, the weight of giant Summer would be 759,840 pounds. That's over seventy full-grown Asian elephants.

And being that big is not good. Not good at all.

BIGGER IS NOT BETTER

In 1926, biologist J. B. S. Haldane published his classic essay "On Being the Right Size," in which he described size differences in the most pleasing way possible:

> You can drop a mouse down a thousand-yard mine shaft; and, on arriving at the bottom, it gets a slight shock and walks away. . . . A rat is killed, a man broken, a horse splashes.

In short, as you get bigger, you have bigger issues with gravity. Or, to put a more modern spin on it: mo' mass, mo' problems.

While the above calculations for Summer show how her weight increases, the question now turns to what translates into all that mass. In mammals, it's largely bones. The strength of bone is proportional to the cross-sectional area, but the cross-sectional area of the bone increases by the square, not the cube.

Upsizing Summer and keeping her in proportion means that her big bones will not be able to support her bigger weight. The extra length won't help; in fact, it will hinder her. The tensile strength of bone depends on the cross-sectional area, and again, that grows by the square.

Think of an elephant, or one of the few remaining rhinoceroses, or a hippopotamus. They're big, and they get around this issue with their legs. Their wide legs.

Wider legs mean the bones inside them have more cross-sectional area, and thus can support the large amount of mass piled on top of them. Even with their wider legs to support them, the big boys and girls of the natural world most often do not move quickly, and when they do, they need to be careful: a simple fall could break a leg. Hippos don't even risk it, and deal with their weight management issue by spending most of their time in water, where their buoyancy decreases the pressure on their bones.

Big Summer retained her basic form (aside from being Clive Barker–ized), and that would mean that her normal human movements would result in more problems. Moving quickly from place to place or jumping isn't happening without breaking bones.

Research by Andrew Biewener has shown that all mammalian bone has about the same strength and maxes out at around 200 megapascals (a unit of pressure roughly equal to 29,000 pounds per square inch). More than that, and the bone breaks. For larger mammals, our larger bones can handle a lot more stress (about five to ten times as much) than we put on them in everyday life and even in extreme sports. But that's normal-sized animals. Start upsizing and the stresses blow through that safety factor and start to break. Every step would mean a broken bone.

As Haldane wrote, when he considered men sixty feet tall (ten times the height of a normal man, so one thousand times as heavy, by volume):

> Every square inch of a giant bone had to support 10 times the weight borne by a square inch of human bone. As the human thigh-bone breaks under about 10 times the human weight, Pope and Pagan would have broken their thighs every time they took a step.

Also, as with the shrunken individual earlier, the internal problems of a giant Summer would be catastrophic. Normal mammalian connective tissue has a range of tensile strength within which it functions. Too much stress and it starts tearing. A shock wave from a bomb or even a loud-enough sound, for instance, could kill you without leaving a mark on you, because it can rip apart your insides, literally.

Since we talked about it when discussing gravity, let's go back to the ocean for another blue whale. Pull one out of the water for an inspection on the beach and you've killed it. The weight of its organs and tissues would crush it. Breathing would be impossible, organs would tear themselves away from the walls of their body cavities, and the heart, as gigantic as it is, would have too much pressure on it to pump blood to where it's needed.

A one-hundred-foot-tall Summer would have the same problem as a beached whale: organs crushing one another, circulatory system unable to move blood for effective gas/nutrient exchange, lungs unable to function.

As with the smaller version of a person, the larger version would have problems with their lungs and gas exchange. To examine this, we need to look at the exchange between oxygen and carbon dioxide at the level of the cells. While gasses are carried by the blood, the actual exchange happens via diffusion, and that's all physics. A simple rule of thumb: you want to have anything that needs to bring oxygen in and get carbon dioxide out be about a millimeter from the gas transport system. In this case, that transport system is the blood. Diffusion of gasses works best over small distances.

Mammals like us handle this exchange via the capillary beds, where artericles, venules, and arteriovenous anastomoses are packed in tight, right near the tissues they supply. Problem solved.

But let's go back to volume. Increase the size of the mammal and you've greatly increased the volume; and thus, you need many more blood vessels to get the blood moving to the cells to drop off oxygen and pick up carbon dioxide. You'd also need a lot more blood, period. That means more red blood cells (and the machinery that makes red blood cells), more plasma, more hemoglobin—more of everything that's in blood.

Gas has a similar problem. Our lungs branch off the trachea into smaller and smaller passageways, finally ending with the alveoli. This smaller and smaller branching serves to increase the surface area at which the gas exchange can occur. The normal, human-but-now-giant lungs of a giant Summer would not have enough surface area to provide adequate gas exchange for her body. In order to survive, giant Summer would need larger lungs, more alveoli or alveoli-like structures, and muscles that would allow for the air to be pulled in and pushed out of those big lungs. Standing near Summer's nose when she inhaled would be like standing in a wind tunnel.

In terms of body heat, remember that Summer's surface area only increased by the square, while her volume (where the heat is made) increased by the cube. She's producing a lot of heat. If she does something that increases her metabolic rate, like throw a temper tantrum, that's just going to create more heat.

In terms of sight: larger eyes let in more light, which would overload the optic nerve. Summer would also have to worry about her rods and cones (human beings' optical light receptors). We have about a half million in each eye, and their size (the receptors are about as big as the wavelength of light) is locked in to their functionality. Make them bigger and you reduce their ability to work.

Additionally, to distinguish between two objects, light from the respective objects must fall on separate rods or cones. If you just make the rods and cones bigger, the larger animal can't see as well as its smaller version, since the incoming light reflected off objects it's looking at falls on a single large cone or rod. It's seen as one large blob. Nature overcomes this by keeping the eyes of large mammals disproportionately smaller than the rest of their bodies. Think of the eyes of rhinos, elephants, and whales; smaller eyes give them a fighting chance, vision-wise.

When it comes to speaking and hearing, take the problems with being small and flip them. Now the vocal cords are much longer and, as such, vibrate at slower frequencies, producing lower tones—which was accurately reflected when giant Summer and giant Beth were talking. In a reverse of their small-person volume, the giants would pass more air over their vocal cords when talking, ultimately producing a collection of loud, low-frequency sounds.

And, finally, going back to strength again: giant, heavy Summer would be weaker than normal Summer, and most certainly would not have any kind of proportionate superstrength. Again, muscle strength is proportionate to cross-sectional area, and while that has increased by the square, the volume, and the mass contained therein, has increased by the cube.

Clearly, being bigger isn't any better than being smaller. Both would end in pretty lousy deaths for the person who sought to take the size-changing ride. Even if we ever get the technology, it won't matter: the physics will get us in the end.

CHAPTER 17

Inventions

★ ★ ✦ ★ ★

Since the very first minutes of the very first episode, *Rick and Morty* has been a showcase for crazy inventions featuring technology that's a blend of real, kind-of-real, and in-no-way-on-God's-green-Earth-C-137-real. While you might think that the bulk of the inventions and technology are just the odd fantasies of the writers' minds, virtually all of them have some kind of basis in real-life technology and modern science.

Let's run through some of the more memorable inventions and technology and see how far behind our real-life inventors are—if at all.

BREATHABLE WATER/FLUID

Rick engineered Froopyland so that its water was completely breathable, so young Beth could never drown.

Sanchez Tech: Rick's breathable water is clear and lovely. It appears to have all the characteristics of regular water.

Real-World Tech: Breathing a fluid is pretty much real, but not water, or at least the water we know. Water in our world doesn't have anywhere near the amount of oxygen to keep us alive, and we lack the mechanisms in our lungs to pull oxygen out of it anyway.

However, it can be done with a type of fluid from a class called perfluorocarbons (PFCs), which are low-surface-tension fluorinated hydrocarbons that are biologically inert, with no color or odor. Experiments in the late 1960s showed that mice submerged in PFCs and that had acclimated to "breathing" the fluid could survive for up to twenty hours, while cats survived for weeks. In humans, total liquid ventilation (TLV) would require a device that would assist with liquid PFC flow into and out of the lungs (essentially serving as a respirator) and would also help with heat and other issues, but the device has not yet been approved.

Partial liquid ventilation (PLV), which fills about 40 percent of a patient's lung volume with PFC, is currently in use to treat acute lung injuries, and has been used to assist premature babies with breathing difficulties. This approach has shown very promising results in both treating and improving the health of damaged lungs.

Additionally, PFCs are used as emergency blood substitutes, since they can carry up to three times the oxygen and four times the carbon dioxide of human blood. There's also no need to lose precious minutes determining the blood type of the patient, as PFCs have no biological factors, so there are no concerns about a blood type mismatch.

Research into TLV is still very active, and along with its potential health uses, investigators are looking at it as an option for deep ocean dives. Due to the tremendous pressures of the deep ocean, divers must use a specialized mixture of gasses for breathing. Fluids such as PFCs barely compress under high pressures, and, as such, TLV could allow deep-sea divers to work without the potentially fatal pressure-related issues. There are still problems to be solved: carbon dioxide removal by PFC at pressure doesn't work experimentally, and the cycling of the fluid through the lungs would require the divers to use a fluid-based respirator. Human lungs are made to breathe air, and our respiratory system is just not strong enough to move the denser, heavier PFC fluid in and out at a rate that would allow for optimum gas exchange.

All that said, that's just in high-pressure environments. Putting your head into a breathable fluid and breathing just to prove a point to your daughter is fully possible.

ANTIMATTER GUN

Rick created an antimatter gun to sell to the assassin Krombopulos Michael, so he could kill the interdimensional alien named "Fart" by Rick.

Sanchez Tech: The antimatter gun Rick designed for K. Michael is gorgeously lethal. It has the requisite unknown substance moving through a chamber, and the sharp points and edges that make it look like it means business. That business is firing antimatter at something that can't be killed with regular matter.

Real-World Tech: Antimatter is real stuff, not just a science fiction go-to when something cool-sounding is needed. And yes, it's exactly what it sounds like: the opposite of normal matter, at the fundamental-particle level. The easiest way to think about it is that it's matter with the electrical charge reversed. For instance, an antimatter electron is called a positron and has a positive charge to the electron's negative one. Antiprotons are protons with a negative charge, while regular protons have a positive charge. In theory, the Big Bang created equal amounts of matter and antimatter in our universe, but to date, researchers haven't been able to find the antimatter, and no one really knows why it's missing.

Just like science fiction has taught you, when antimatter encounters regular matter, the two annihilate each other and release energy. While this is certainly the plan K. Michael had for Fart, matter-antimatter annihilation is far more common than that. Images of the brain produced by Positron Emission Tomography

(PET) are possible thanks to radioactive substances injected into the brain that emit low-energy positrons. When the positrons react with the matter of the brain, a small burst of gamma radiation is produced (don't worry, the amount released is harmless). The imager looks for that radiation, and from that can produce its detailed images of the brain.

Ongoing research is looking at antimatter-matter reactions as a possible fuel for moving through space, and the US Air Force has been interested in making antimatter weapons for over a decade, though both processes are quite complicated. Chief among the complications is storing the stuff. Since antimatter will annihilate the regular matter of your container, magnetic field "jars" are often used, which are able to suspend the antimatter without it meeting matter.

Also, the price to create and collect antimatter runs about a billion dollars per milligram. Again, as with virtually all the other "real" tech that Rick works with, his understanding of it and ability to manipulate it is far greater than ours, especially if he can produce large quantities of antimatter to use as a weapon.

ARMORED COMBAT SUITS

Upon realizing that the entire community they were in was going through a purge night, Rick and Morty called for some help in the form of fully armored exoskeletons, which allowed them to cut through the villagers like they were made of butter.

Sanchez Tech: Rick made these fully armored suits with jet boots and tons of weapons as a fail-safe in case of multiple contingency scenarios, and what he and Morty were experiencing on the purge planet certainly qualified. The suits were constructed over the wearer, so no time was lost getting into them.

Real-World Tech: Mechanically assisted movement is a wide-open field of research and engineering, with three distinct paths:

in medicine, assisting individuals who are injured or otherwise have difficulty walking; in business, helping workers in factories who suffer from fatigue and injury from work; and, of course, in the military, preventing injuries and increasing performance. Dozens of companies, universities, and agencies are in on the race to design a usable suit for soldiers that would function in action on the battlefield and help them with lifting, loading, and work at a base camp when not in combat.

One of the more publicized military-grade projects is for the Tactical Assault Light Operator Suit (TALOS). President Obama spoke about it at a briefing in 2014 when he said that the US military was "building Iron Man." The suit is a full exoskeleton that will be bulletproof and weaponized, have sensors that can monitor the wearer's vital signs, and provide the wearer with enhanced strength and perception. The suit has had a few setbacks, most notably the issue of providing enough power for it while keeping it lightweight.

But TALOS is not the only military-level exoskeleton in development. DARPA has an entire program dedicated to exoskeleton research and design. Longtime US military defense contractor Raytheon has developed its own suit line, the XOS, and has shown working prototypes that allow for the transport of heavy items. Raytheon promises that a combat-variant XOS suit is in the works as well. Their competitor, Lockheed Martin, has also shown off its own ONYX device, which is to be worn on the legs to aid with difficult terrain or tasks that would otherwise strain the knees and legs.

The suits worn by Rick and Morty are just slightly ahead of the curve when it comes to size and weapon systems, although developers are catching up. After proof-of-concept models have been made by various research and engineering teams, the focus shifts to making the devices smaller and slimmer, so they can be easily worn by soldiers on the battlefield.

Rick's suits seem to have gotten around the hurdles that still trouble modern exoskeleton designers: providing enough power while keeping the suit lightweight, keeping it comfortable enough to wear for hours at a time, and keeping it quiet. Loaded down with machinery and bearings, many exoskeletons are plagued by engine or motor noise.

Unfortunately, none of the proposed suits have jet boots like Rick's. Yet.

GRAPPLING SHOES

Seeing that they need to climb down a cliff, Rick gives Morty grappling boots that will allow him to walk down the cliff's face. Morty forgets to turn them on and falls to the bottom, breaking his legs in the process.

Sanchez Tech: While looking like ski boots, Rick's grappling shoes allow you to walk on virtually any surface, as long as you remember to turn them on first. Significantly, as Morty learned the hard way, they need power to work.

Real-World Tech: Rick's shoes are awfully similar to the pads on a gecko's feet, which allow it to climb up smooth surfaces as well as it can crawl along horizontal ones. Geckos can do this because the surface of each pad branches and branches, finally becoming microscopic projections called setae. At this scale, the setae use the natural electrostatic forces, called Van der Waals forces, to "stick" to the wall, allowing the lizard to climb straight up.

This ability has been mimicked by researchers using microscopic silicon rubber bristles that behave just like the setae on a gecko's foot pad. Using some clever engineering, teams at Stanford and the Department of Defense have developed devices that allow a human to climb up a glass surface, pretty much like a gecko does: one adhesive pad is placed against the surface and weight is applied; then the pad on the other hand or foot is placed against the surface, and the weight is shifted. To remove the pad, all the climber must do is remove their weight from it, just like a gecko does.

Since it's the close positioning of the setae, both real and artificial, against the surface that creates the attraction and subsequent climbing ability, this approach wouldn't allow you to climb on *any* surface, like Rick's boots do, but it's a pretty close match to Rick's tech. And there's no on/off switch to remember.

WRISTWATCH LASER/PARTICLE BEAM

Rick has been shown to wear anywhere from three to five watches, and we've seen one used to vaporize the general in "Get Schwifty" and another used to slice through secret service agents in "The Rickchurian Mortydate."

Sanchez Tech: Rick wears a lot of watches, and many of them have special features, including a satellite dish, a personal shield generator, a hologram clone generator, a communicator, and, of course, a watch. But two watches appear to be straightforward weapons: one is a particle beam generator, and the other can emit a powerful laser beam.

Real-World Tech: Both of Rick's watches use forms of directed-energy weapons, and we've got those. Weapons such as these damage the target with focused energy that can include lasers (light-based), microwave beams, or particle beams (energized subatomic particles). Particle beam weapons use electromagnetic fields to accelerate charged particles to near light speed, then shoot them at a target. The damage to the target is a result of the extremely high kinetic energy of the particles in the beam transferring to the object, which instantaneously superheats the surface and can ionize the deeper layers. All of this is particularly harmful if the target is electronic or biological in nature.

A laser (not sure if you knew this, but the word "laser" is an acronym: Light Amplification by Stimulated Emission of Radiation) is produced when light is shined into a specific material, such as a ruby, held in a chamber with reflective walls. The incoming light causes the release of photons inside the ruby. Single photons bounce back and forth between mirrored surfaces, actively causing other photons with the same wavelength and direction to be emitted from other atoms of the material. Ultimately, these organized, coherent photons are shot from the laser in the form of a beam. High-powered lasers instantly heat the surface of the target and can cause it to sublimate (turn from a solid to a gas).

Both technologies are under active research by the military as weapons, and prototypes have been rolled out, but there is a limitation: power. The beams are only as powerful as their energy sources. If you want a laser to do more damage, you must pump more power into it, and power sources come in two varieties: batteries and power plants. Portable lasers or particle beam weapons powerful enough to cut through or vaporize humans would require a tremendous amount of power for that killing shot, and unless the power source was self-regenerating or infinite, the weapon would be good for only one shot. Currently, weaponized lasers and particle beams are platform-mounted on ships or trucks that can support their power demands.

All this is not to say that prototypes of wrist-mounted lasers don't exist. Hobbyists have been cranking them out for a while now, and their YouTube videos show some pretty impressive models that can light matches and scorch wood at a distance. But when it comes to lasers and particle beam weapons in our world, always remember on thing: no matter what science fiction has taught you, the beams themselves are invisible in air.

OVENLESS BROWNIES

As Doofus Rick and Jerry bonded while the Smiths' house was occupied by agents from the Trans-Dimensional Council of Ricks, Doofus Rick showed Jerry how to make ovenless brownies using chemicals.

Sanchez Tech: Along with some other, unnamed ingredients, Doofus Rick's ovenless brownies were made with titanium nitrate and chlorified tartarate.

Real-World Tech: The good news: ovenless brownies exist. The better news: you don't have to use any kind of weird or possibly toxic chemicals to make them.

With baked brownies, you've got plenty of opportunity to add moisture, including butter, brown sugar, and eggs. It's the interaction of those ingredients that produces gas bubbles in the brownies as they cook, giving them a slightly cakey appearance and texture. The heat is also supposed to kill the bacteria in the eggs if you've got a bad batch—Salmonella does not add anything to brownies.

With ovenless or "no-bake" brownies, you won't get a cake-like texture, and your ingredients must do multiple jobs. Any liquid needs to carry flavor. Any dry ingredient needs to act as a binder and add the structure and support that baked brownies would have, while also absorbing the liquid so everything sticks together. And the melted chocolate will need to partially resolidify when mixed with the other ingredients, adding more of the structure that baking provides. But a recipe exists:

* Sweetened condensed milk (about 14 ounces)—your liquid and sweetness.
* Unsweetened or bittersweet chocolate (2 to 4 ounces)—you've already got the sweetened condensed milk, so you

really don't need to add more sweetness with the chocolate; you just need the chocolate flavor.

* Graham cracker crumbs (about 2 to 2.5 cups)—this is your binder and what will soak up the liquid.
* Chopped nuts (about 0.5 to 1 cup)—these add just a little oil to the moisture side of things, and will add body and flavor to the brownies. Use walnuts, hazelnuts, or whatever you prefer.

Spray a pan with cooking spray, or line it with parchment paper if you're fancy. Mix your dry ingredients together, reserving about half the nuts for the top of the brownies when they're done. Heat the condensed milk and chocolate together in a saucepan over low heat, until the chocolate has melted. Once the combination is melted—and it should be thick—blend in the dry mixture. Sprinkle the remaining nuts on the top and chill the brownies in the refrigerator for three to four hours.

No titanium nitrate needed.

NANOTECHNOLOGY—MORPHIZER-XE AND MORTY'S NANOTECH VIRUS

Both elements hint at the advanced use of nanotechnology. In "The Whirly Dirly Conspiracy," the Morphizer-XE (which Rick may or may not have built) can change one type of matter into another, and in "Rest and Ricklaxation," "good" Rick shoots Toxic Morty with a bullet laced with an encrypted nanobotic virus.

Sanchez Tech: If Rick created the Morphizer-XE, with its tiny aliens inside running it, that and his infecting Toxic Rick's Morty with a nanobotic virus show that our mad scientist is able to create and alter things on the nano scale.

Real-World Tech: Nanotechnology is one of the fastest-growing fields of research in the world. The prefix "nano" specifically means "billionth," as a nanometer is a billionth of a meter. The term, however, is most often used to cover different scales, mostly in the near-atomic and molecular range, and the technology side refers to the engineering of structures, devices, and other systems of that size. Nanoparticles and nanomaterials are just things that are on the scale of 1 to 100 nanometers in length.

One of the promises of nanotechnology is that it could be used to create functional robots at that scale. These robots, as those on both the pro and con sides of the technology have claimed, could be programmed with simple commands, such as to kill cancer cells inside a body. But, as some researchers and ethicists have pointed out, if the robots could be programmed to find cancer cells among the myriad of cell types in the human body and kill them, they could also be programmed to target specific segments of DNA and disrupt them as well, in essence killing an extremely specific target or group of individuals who share a common genetic trait. The specific DNA-targeting of a nanobot virus would be what Rick was talking about with the virus that was going to kill Toxic Morty (although the Toxic version of Morty would share the same DNA as de-toxed Morty), while programming nanobots to seek out and kill a certain race- or ethno-specific gene pattern would be the recipe for targeted, ultraspecific genocide.

Our current technology is years and years away from that, and significant questions remain in regard to robots being able to assemble themselves out of atoms, over and over, with zero errors. However, that doesn't stop the term "nanobot" from showing up in the news, usually attached to a medical or technological claim. Most often these "nanobots" are hybrids that incorporate nanomaterials with small-scale organisms rather than robots. One such example is a virus called a bacteriophage, with slightly modified genetic material and nanoscale magnetic particles attached to its outer surface. The genetic alteration makes the bacteriophage seek out and bind to specific, live, and viable bacteria. Then the magnetic material can be used as a signal to indicate the presence of the bacteria in the sample, or the bacteria can even be gathered up by using a magnetic field. When used to detect bacteria in a sample, this approach reduces the time needed for culturing the bacteria and growing it from days to just hours.

Another of nanotechnology's potentialities is nanoscale machines that would work with individual atoms to transmute one atom into another. Alchemy's promise of obtaining gold from base elements might finally be realized, as our smallest machines could go into an atom's nucleus and adjust the number of protons and neutrons, possibly doing something like changing a statue of a garden gnome from plaster into diamond with just the twist of a dial on your Morphizer-XE.

Unfortunately, this use of the Morphizer-XE is as unlikely as its other use, which grew Summer and Beth to monstrous proportions (see the chapter on Growing and Shrinking People). At the nuclear level, the force holding neutrons and protons together is called the strong nuclear force, which is the strongest of the four fundamental forces of the universe. Acting at a very short distance—the space between adjacent protons and neutrons in the nucleus—this force is 100 times stronger than the electrostatic force, 10,000 times stronger than the weak nuclear force, and 10^{38} times stronger than gravity. Having a nano- or smaller-scaled machine go into a nucleus and pluck out or add protons and neutrons just isn't possible.

That's not to say that we can't change one element into another; that already happens in nuclear fission and nuclear fusion, as well as in radioactive decay. Respectively, this is when the nuclei of atoms split apart, smash into one another and stick, or just spontaneously fall apart. In theory, and in practice, nuclear physicists have realized the alchemist's dream and transmuted one element into another. The thing is, the energy needed to change one atom into another is often massive, and nothing you'd want to be around. Gold, for example, can be produced only via the transmutation that happens in a supernova, when a certain type of star reaches the end of its life and blasts virtually all its mass out into the space around it in a violent explosion.

Nanotechnology today isn't about engineered viruses or transmutation as much as it is about creating new materials and improving upon existing ones. Rick's applications are, of course, way past where we're at.

NEUTRINO BOMB

Rick occasionally creates neutrino bombs that will, in his words, kill everyone on Earth (or whatever planet they happen to be on). Luckily, by the time Rick uses such a bomb with the Vindicators, Morty has disarmed so many that it's not that much of a threat.

Sanchez Tech: Thankfully, we've never seen one of Rick's neutrino bombs go off, but from their appearance, they look like high-tech devices blended with the traditional "bomb" elements of movies and television. But as Morty told the Vindicators, there's a 40 percent chance that any given neutrino bomb built by Rick is a dud.

Real-World Tech: Neutrinos are real. Neutrino bombs are not.

Neutrinos are subatomic particles with no electric charge and very little mass that barely interact with ordinary matter via the weak nuclear force. Add those things together and that makes neutrinos incredibly hard to detect. They carry energy and are produced naturally by the nuclear decay of radioactive elements, as well as by the sun and other stars. Roughly 2 percent of the sun's energy is carried by neutrinos.

Think about those two things and then look at your hand. About a trillion neutrinos from the sun passed through it in the second it took you to look at it. A trillion more passed through while you moved your eyes back here to read this sentence. Another trillion right now. You didn't (and won't) feel anything, because, given their aversion to interaction, one neutrino in, say, a dozen years will hit the nucleus of one of your body's atoms, if conditions are just right. If a neutrino does hit a nucleus, it could hit a neutron, turning it into a proton, while causing the former-neutron-now-proton to spit out an electron. This is called beta radiation and can be harmful in high-enough doses. But again, most neutrinos pass through the Earth (and all its creatures) without even slowing down.

There's your problem with a neutrino bomb: it would be non-lethal, since most of the particles would whiz right through the intended victims, as well as the planet upon which they were standing. If you want to fix that problem—like Rick may have done—you've got an issue of magnitude. You would need to dramatically increase the number of neutrinos to up the chances that enough of them would interact with enough matter to have a lethal effect on not just one but all the people on the planet. As physicist Dr. Martin Archer calculates, that would be roughly 5.0×10^{15} neutrino collisions with nuclei per person. Unfortunately, only about 1 neutrino in 10^{21} neutrinos react with matter, so your number of total neutrinos for a lethal dose per person rockets up to 10^{36} neutrinos per square meter per second. Multiply that by every person on the planet, and then up the numbers some more because neutrinos released from the bomb will spread out as they travel, and you're looking at somewhere over 10^{52} (10,000,000, 000,000,000,000,000,000,000,000,000,000,000,000,000,000,000) neutrinos per square meter per second to get the job done and kill everyone on Earth. Just for a point of reference, it would take our sun 2.5 million years to produce that many neutrinos.

Also for consideration, if you're talking about a fusion reaction in a star, only around 3 percent of the energy of that reaction goes into the neutrino itself. Creation of neutrinos at this energy level is just horribly inefficient. To create more neutrinos, you'd need a bigger reaction; something on the scale of a supernova, where the stellar core collapses to become a neutron star, should do it. And it seems a little insane that, with all the energy such a supernova would produce, Rick would only want to skim off the neutrinos for his bomb.

But, as always, this is Rick. Who knows?

SELF-AWARE ROBOTS

Rick can create robots that, along with their artificial intelligence, are aware of who and what they are. This tends to lead to difficulties. Over and over.

Sanchez Tech: Rick's robots contain a level of artificial intelligence that can make them indistinguishable from the real people they're copying, which is one thing, but he perhaps goes too far when programming their intelligence. That is, they reach a point where they begin to wonder who they are and what their purpose in life is.

Real-World Tech: Artificial intelligences surround us in a multitude of forms: some are painfully obvious, emphasizing the artificial over the intelligence, such as Siri and Alexa and their questionable—but improving—spoken-word recognition. Others are completely invisible, seamlessly integrated into daily life, such as what shows up on our social media feeds or when Google sorts your photos into groups by the main object in the picture. They are already smarter than us and are getting better at their jobs by leaps and bounds, but by many metrics, they are not more intelligent than we are. That is, they can hold and access more information ("smarter") but can't organize it and act upon it in ways that are inherently human.

Artificial intelligences can imitate intelligence, and we often project intelligence on them, with the end result that we trick ourselves into thinking that they're more intelligent than they actually are. Alexa's algorithm may play songs by the artist you request, but if it plays "your" song on a specific date, that's just a coincidence. Alexa doesn't know about the date or its significance. If you doubt that we project more intelligence and greater abilities onto our digital assistants than they possess, keep track of how many times you talk back to one.

Sure, they can be creepy, but that's because they're doing the best they can, given their programming. What we tend to label as novel quirks of behavior that perhaps point toward true intelligence or self-realization are almost always solutions to problems presented in a way we hadn't anticipated. They're terrific problem solvers, and will only get better.

But are they self-aware? This would be a new path for AIs. They would have to develop consciousness, and even today, with all we know about psychology and the human brain, we know precious little about human consciousness.

What we do know, and perhaps using this as a stand-in recipe for human consciousness, is that a self-aware robot would require a body to gather and respond to stimuli, a mind to integrate and analyze the stimuli and use that information to synthesize new ideas, and a means of communication. But even with these three pieces, there's no guarantee that self-awareness will bubble up and the robot will one day spontaneously "awaken" and recognize itself as an individual being, independent of others and in control of its own mind, body, and destiny. After all, we have rudimentary forms of robots with all the abilities mentioned above, and not one has yet asked "Why am I here?" or "What is my purpose?" on its own. And even if one did, it would be unclear if it was truly conscious and self-aware or if, as with other behaviors, it's only imitating being such due to a unique solution path through its programming.

In short, Rick's robots, with their self-awareness, are far, far beyond what our most advanced AI is now. If Rick is creating robots that are self-aware, he's creating a new form of life.

TELEPORTATION

Hijacking the teleportation mechanism (the one with buttons and dials) for his own uses, Rick teleported the Citadel into a Galactic Federation Prison, causing huge amounts of death and destruction, in "The Rickshank Rickdemption," and the Cromulons teleported the entire Earth to a new location for it to take part in *Planet Music* in "Get Schwifty." Also, Rick sent instructions to NASA on how to build a teleporter when he and Morty were trapped in the menagerie in "Morty's Mind Blowers," and Stacy was accidentally teleported into the toxic world inside the tank of the detoxifier in "Rest and Ricklaxation."

Sanchez Tech: We're separating teleportation from Rick's portal technology, which we're assuming is wormhole-based. Let's define teleportation as changing position instantly and without the use of an entry or exit point. Blink in, blink out. With that said, Rick prefers to travel via portal rather than teleportation, but as "Morty's Mind Blowers" showed, he does know about teleportation and how to use it.

Real-World Tech: Strictly defined, we're talking about moving from point A to point B without traveling through the space between the two points. Teleportation was first proposed as a legitimate scientific theory in 1993 as something allowable within the world of quantum physics. As the initial theories explained, teleportation should be possible at the quantum level, but one problem would be that the object at point A would be destroyed, while a copy of the object would appear at point B.

In 1998, a group at CalTech proved teleportation could work, at least for photons. Their teleportation method utilized quantum entanglement. In this phenomenon, a pair of particles are created at the same time and place. Effectively, the two particles act and respond the same to change. Even if the particles are separated, they still demonstrate this close relationship: if one particle changes, the other changes, no matter how far apart they are. Now, bring in a third particle and have it interact with one of the first two particles. The interaction with the third will destroy the first particle, but the other (second) entangled particle will reflect that interaction and change, effectively becoming the first particle. Essentially, the CalTech experiment managed to teleport a single photon about one meter.

More demonstrations have followed that have teleported photons, laser beams, and information across varying distances, up to about 100 km. The record holders are Chinese researchers who, in an experiment using entangled photons, teleported them to a satellite in low-Earth orbit—roughly 1,400 km away—in 2017.

Our teleportation technology is locked in at the photon level, and probably will be for a long, long time. In order to transport a human, following what we know from experiments with photons, information from every atom in the body (that's around 10^{28} atoms) would have to be analyzed. Then, billions of entangled pairs of particles would have to be established at the destination, and then you could send the original information. But, as with photons, the analysis and sending would destroy the original atoms—and there would be no room for error at the new location. The computing power required to analyze, entangle, and send a human is staggering. The Citadel? Unfathomable.

If Rick has teleportation figured out, he's years and years ahead of our physics.

INVISIBILITY

During the final fight between the president and Rick (and most likely before that), the president's Invisi-Troopers had Rick surrounded. Rick could, he said, see them the whole time.

Sanchez Tech: The soldiers' invisibility wasn't Rick's invention, but given the disdain he had for it, it's easy to believe that Rick has created similar but far superior technology in the past.

Real-World Tech: Invisibility is a hot field of research in modern technology, partly fueled by a certain cloak worn by a certain teenage wizard, partly fueled by the growth of metamaterials, and partly fueled by the collective desire of the militaries of many nations.

The metamaterial approach is to cloak the object with a specialized material that will allow light—and, ideally, all electromagnetic energy—to pass around it as if it were not there. Metamaterials are constructed to react to electromagnetic radiation in ways that are not found in nature. In the case of classic invisibility, light would not interact with the metamaterial, and an observer would see only what is on the other side of the cloaked object no matter their angle, rendering the covered item invisible. Metamaterial cloaking in infrared, microwave, and radio regions of the electromagnetic spectrum has already been developed and will become a common feature of military craft from drones to next-generation stealth jets.

While a metamaterial cloak for visible light has been slower in coming, new developments have demonstrated an ultrathin cloak that can wrap around an irregularly shaped object and reflect light back as though the light were hitting a flat mirror. So far, the research has been performed only on microscopic objects, but it may soon be able to scale up.

There is also a digital method of invisibility outside the field of metamaterials. In this approach, the object to be cloaked has numerous cameras attached to it that capture images around the object from all angles. These images are then displayed on the opposite side of the object in real time, making the object appear invisible. The amount of processing power needed to make the images continuous and mapped in their precise locations is immense, and this is best used with larger objects with regular geometry.

A simpler technique than this has been developed at the University of Rochester, where four lenses are used in a certain configuration. This setup manipulates light in such a way as to render any object invisible when viewed through the lenses, allowing for continuous, multidirectional cloaking without any distortion of the background. The Rochester team has also created a digital version with a wider field of view that allows for a changing viewpoint.

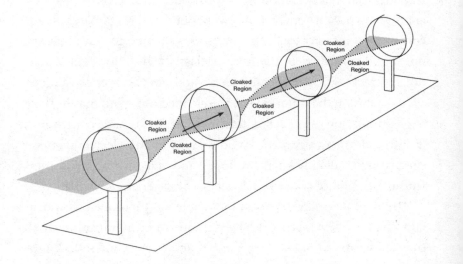

Despite invisibility's use by the president in *Rick and Morty*, as well as mentions of military uses here, much of the research on metamaterials and the optics they manipulate is aimed at commercial applications, such as video displays that can show flat images on irregular surfaces without distortion, imaging in medical science (for instance, providing image assistance during surgeries), and added safety in transportation by making parts of vehicles "invisible."

But Rick will be able to see it all anyway.

HANDHELD BEAM/LASER WEAPONS

Rick has a sidearm that fires energy beams and is battery-based, but that's not the only handheld gun that's showed up in the series. The Gromflomites use rifles and sidearms specific to the Galactic Federation, and the Ricks of the Citadel as well as SEAL Team Rick have a variety of such weapons.

Sanchez Tech: Rick's laser gun is a handheld directed energy weapon that works within a set power range that yields a variety of outcomes, from laser beam to lightning bolt. Other handheld weapons show similar abilities and characteristics, including pulse settings for kill and nonlethal incapacitation.

Real-World Tech: Again, as with Rick's wristwatch-mounted laser, beam technology does exist in our world. Lasers produce light that is all the same wavelength, in-phase, coherent, and synchronous. While we looked at simple laser emissions with Rick's watch, that's not the end of the story for directed-energy weapon technology in our world. Some ideas floating around labs and research centers today sound like they came straight from Rick's garage.

In the early 2000s, the US military was actively researching and developing a pulsed energy projectile (PEP) weapon. When fired, the weapon would emit a burst of infrared laser light, the first part of which would create a small amount of plasma upon contact with the target's surface. The second part of the pulse would then energize the electrons of the plasma to the point where it would explode. The explosion would create a pressure wave, which would be of a large-enough magnitude to knock a normal person down, as well leaving them in extreme pain due to the electromagnetic radiation affecting their nerve cells. The weapon was designed mostly as a nonlethal method of riot control, and reportedly had a range of up to 2 km. While the military promoted the weapon as such, critics have pointed out that, as with other energy weapons, increased power results in increased effect at the target—to the point where it could kill.

In case you'd rather dazzle your target before incapacitating it, the US Army has developed another weapon called the Laser-Induced Plasma Channel (LIPC), which shoots lightning bolts. Seriously. This is accomplished by firing a high-powered laser beam for a very short time (about two trillionths of a second). The energy of the laser is enough to pull electrons off air molecules, creating plasma along the beam's path. That plasma channel acts like a filament and conducts electricity much better than regular air. The weapon then releases a high-voltage electrical charge. The lightning bolt travels down the channel of plasma until it gets near something that will offer it a lower-resistance pathway to the ground, whether that something is a person, a car, or other object that would allow high-voltage electricity to travel through it. If you want to think of it as a lightning gun, go right ahead.

Both weapons have, at their hearts, the same issue with power mentioned earlier, and prototypes for both are, or will have to be, mounted on vehicles to carry the power supply. Every country with a military is trying to scale down the size of their directed-energy weapons, and China may have already won that race with the ZKZM-500. The six-pound weapon reportedly has a range of over 2 km and carries enough power for one thousand two-second shots. The device is intended to be used as a non-lethal weapon, but again, increased power means an increased result at the target.

So we're not at the stage of the guns of *Rick and Morty* just yet, but give us a couple of decades and we just might be there.

FREEZE RAY

In the following cases, a freeze ray was used by Rick to stop someone in a nonlethal manner: the bully Frank Palicky in the pilot episode, Jerry in "Close Rick-Counters of the Rick Kind," and to freeze C-137's Jerry, Summer, and Beth in "The Rickshank Rick-demption." While Frank fell over and tragically shattered, Rick C-137 requested that Jerry be thawed out as he was being taken to the Citadel.

Sanchez Tech: A simple, small device, the freeze ray freezes its target solid, as evidenced by Frank's shattering when he tipped over and smashed on the floor in the pilot episode.

Real-World Tech: For such a staple in science fiction, as well as in *Rick and Morty*, a freeze ray is something that, to be technical, doesn't make sense. That's because of what "cold" is. Cold isn't a thing in and of itself. Cold is a lack of heat. Heat is what moves from one object (the warmer one) to another (the cooler one). If something is colder than something else, it just has less heat than the other thing, not more cold.

That said, virtually the only way to make something cold is to take heat away. When heat is taken away from an object, its molecules and atoms move more slowly. Take more heat away, the movement is slower still. If you take enough heat away, the molecules will change into their next-lower state of matter. For instance, taking heat away from a liquid would make it a solid.

Rick's freeze ray has been shown to project an energy that en-gulfs the target and then freezes solid. Maybe the ray shoots a substance that completely absorbs the heat of the target and the water vapor in the surrounding area, which would explain the thin casing of ice around the victims, but like many of Rick's inventions, that's beyond the scope of the physics we know.

MOON TOWER

Toxic Rick finds that the Moon Tower on the edge of town is the perfect place to both amplify and beam out toxic energies, creat-ing a toxic Earth.

Sanchez Tech: Rick didn't invent the Moon Tower, but his toxic counterpart finds it perfect for his intents.

Real-World Tech: Moon Towers aren't new technology but actually a nice nod back to some cool nineteenth-century technology. Moonlight Towers were used in several towns in the nineteenth century in the United States and Europe. Over a hundred feet tall, the towers were most often outfitted with carbon-arc lamps, which provided extremely bright light to illuminate the surrounding areas, sometimes reaching several blocks in diameter.

The towers were often used when gas street lamps were impractical or the expense of individual lamps, whether gas or incandescent bulbs, could not be justified. Slowly, incandescent street lamps increased in popularity and use, and one after another, the Moon Towers were dismantled. Today, only Austin, Texas, still has operating Moon Towers, fifteen in all. They are kept in their original, functioning condition and are a cherished part of the city's history and landscape.

TOTALLY MADE-UP STUFF

In "The Ricks Must Be Crazy," Morty's attempt to come up with a science-sounding name for a car part gets a classic retort from Rick: "Quantum carburetor? Jesus, Morty, you can't just add a sci-fi word to a car word and hope it means something." While Rick may warn Morty away from such things, at least a couple of Rick's devices fall into this category: "Rick Potion #9"'s ionic defibulizer and "Raising Gazorpazorp"'s quantum resonator. The names of each item keep them squarely in the world of Rick's tech, and not that of our world. Sometimes it *is* pretty fun to take sci-fi words and put them together with normal words and pretend they mean something.

Also, hopefully none of the tech that Rick, the ever-loving dad, made for Beth is real. This includes: a whip that forces people to like her, invisibility cuffs, a literal parent trap, a teddy bear with anatomically correct innards, location-tracking stitches, rainbow-colored duct tape, night-vision googly-eye glasses, sound-erasing sneakers, false fingertips, fall-asleep darts, a lie-detecting doll, an indestructible baseball bat, a Taser shaped like a ladybug, a fake police badge, mind-control hair clips, poison gum, and a pink, sentient switchblade.

If they are, though, they shouldn't be given to children. Get help, Beth.

CHAPTER 18

Are You Living in a Simulation Operating at 5% Capacity?

★ ★ ✦ ★ ★

Are you reading this book right now? Are you reading this book right now, on Earth, a planet in the Milky Way galaxy in the Laniakea supercluster?

Or rather, are "you" nothing more than some lines of what passes for "code" in a simulation run by a hugely advanced society, on something that vaguely resembles a computer but is exponentially more powerful and complex? Is the world as you know it being created solely for you, with nothing existing beyond your immediate visual field?

How can you prove either one of those? Maybe by opening up a rat and finding some seriously sloppy craftsmanship, or asking your son-in-law (who may or may not be real himself) to take off all his clothes and fold himself twelve times. At least those approaches worked for Rick.

That's simulation theory. The idea that reality—everything our senses experience, every thought or feeling in our brain—is all simulated, and nothing is really real. Or maybe something of it is real, or was real once but isn't anymore. How would we ever know? And if we learned that it was a simulation, would we be able to find out anything about whoever (or whatever) was running it?

These are the types of questions that, when you ask them out loud, make you sound like you're stoned. So, naturally, they've shown up in *Rick and Morty*. Three times, actually: in "M. Night Shaym-Aliens!," "Mortynight Run" (as the game "Roy" in Blips and Chitz), and "The Rickshank Rickdemption."

Simulation theory's appearances in the first three seasons of *Rick and Morty* reflect its popularity with the audience.

WHEN DID WE START WONDERING?

As a concept, simulation theory has comfortably lived in the realm of philosophy since it was first thought up, but in recent decades it has started to push its way into science. And both sides of the issue—that we are living in a simulation, and that we're not—have equally strong advocates, with equally strong arguments.

The idea of being in a computer-based simulation is relatively recent, but the idea of life as we know it being a dream began as soon as we could settle down and think for a few minutes instead of having to spend all day hunting for food and fighting off neighboring tribes. As it does today, analyzing the very nature of reality can get a little trippy in a hurry, and led our ancestors to some . . . interesting ideas. However, just like us, our existentialist predecessors were working with the best tech they had to analyze the universe around them.

One of the earliest mentions of what could be interpreted as simulation theory comes from Zhuangzi, a Chinese philosopher who lived around 300 BCE. According to his writings, Zhuangzi once had a dream that he was a butterfly, enjoying his butterfly life to the fullest. Upon waking, though, he had a thought—was he now a man who had been dreaming that he was a butterfly, or was he a butterfly who was now dreaming he was a man?

A famous Western contemporary of Zhuangzi—a guy you may have heard of named Plato—had similar theories, that what we know as life may be only a mere sliver of a larger world that we can't see. In *The Republic*, Plato presents us with "The Allegory of the Cave," where a band of prisoners has been chained facing a wall in the cave since birth. The prisoners know nothing of the world outside—all they know of anything other than themselves are the shadows on the wall made by people passing by the fire behind them. The prisoners come to believe that the shadows they see are things as they actually are. The prisoners are like us—what they know is limited by their perception. If all they know are shadows on the wall, that's their reality. When a prisoner is taken out of the cave and shown the larger world, they can barely comprehend it and long to return to the cave, where things made sense.

Plato and Zhuangzi's thoughts have lasted for nearly 2,500 years because they continue to resonate. They hit every button we humans like to push: What is real? What does "real" even mean? Is there someone (or something) controlling everything? Are we being watched as an experiment? Or entertainment?

I told you this is an idea that has been more at home in philosophy than science. Don't worry, though—the science is coming up.

SOME DEFINITIONS AND LIMITATIONS

Before we get too deep into the weeds, we need to set some ground rules for what makes up simulation theory. This is not the multiverse we're talking about here—even though a simulation can contain a simulated multiverse. Simulation theory isn't about world after world after world that are all very similar except for small variations—unless we're talking about various, different simulations . . . which brings us to the first warning. It is extremely easy to go down a rabbit hole or get into infinite regression with simulation theory. This is perfectly captured in an unaccredited anecdote from 1838 about a young schoolboy and an old woman living in the woods:

"The world, marm," said I, anxious to display my acquired knowledge, "is not exactly round, but resembles in shape a flattened orange; and it turns on its axis once in twenty-four hours."

"Well, I don't know anything about its axes," replied she, "but I know it don't turn round, for if it did we'd be all tumbled off; and as to its being round, any one can see it's a square piece of ground, standing on a rock!"

"Standing on a rock! But upon what does that stand?"

"Why, on another, to be sure!"

"But what supports the last?"

"Lud! Child, how stupid you are! There's rocks all the way down!"

In the twentieth century, "rocks all the way down" was changed to "turtles" because the old woman's story was altered to have her explain that the Earth's crust sat on the back of a giant turtle, and that turtle sat on the back of another turtle, and so on . . . all the way down.

Discussions of simulation theory can rapidly degenerate into "turtles all the way down," or, rather, infinite digressions about infinite regressions, leaving minds spinning and everyone feeling like there may have been something in their drink. Let's try to stay on the surface—or at least within sight of the top turtle—as we go along.

What is simulation theory? Simply put, it's that we're living in a reality that's simulated—that is, not real. For example, in the *Twilight Zone* episode "Stopover in a Quiet Town," a couple wakes up in an unfamiliar town where the houses and all the world are props, and find that they are the pets of a little alien child. A similar, more recent example would be the film *The Truman Show*—but with real, physical, or "analog" versions of the digital examples we see today in popular culture.

In simulation theory, the believable reality is created for people in the simulation by a far superior technology. This can take the form of simulation bays on the Zigerion spaceship, with nanobots rapidly creating the "world" with which Rick and Jerry interact, or more direct, brain-only simulation, like in the game of "Roy" and the interrogation used by the Federation.

In theory, time for the individual living in the simulation does not have to be moving at the same rate as that for the individual running the simulator, and often moves faster than the time the simulator experiences. Within that time frame, the simulated individuals are free to do whatever they want, within the programming or the rules of the simulation. And this includes exploration, discovery, living full lives, and even developing their own technology, to the point that they can make their own simulations.

As mentioned, things can even get pretty recursive pretty fast—you could get deep down in a stack of simulations, exponentially far from the original, or root, simulation; in other words, you may be reading this book in a simulation that's nested within another simulation that's nested within another simulation, and on and on and on until you feel less than insignificant.

Around the mid-twentieth century, the idea of simulation theory took a slow, stumbling step toward a more science-y interpretation, as science fiction writers started to understand that we were soon going to have ways to make our dreams reality.

Some early versions of simulation theory can be found in science fiction stories like Daniel F. Galouye's *Simulacron-3* (1964), in which individuals were computer programs that lived in a virtual city created and run for market research (the book was turned into a German made-for-TV movie in 1973). Philip K. Dick questioned the nature of reality many times in his stories, but nailed the idea of simulation theory in 1966's "We Can Remember It for You Wholesale"—the basis of which would be used for the *Total Recall* films of 1990 and 2012. *Doctor Who* showed a simulated world—ironically called "the Matrix"—in 1976's "The Deadly Assassin," and by 1987, science fiction TV audiences were getting a regular dose of the concept in *Star Trek: The Next Generation*'s Holodeck.

And then, in 1999, *The Matrix* landed in mainstream popular culture, and brought with it the concept of a simulated reality. Everyone knew of the postapocalyptic world of Morpheus, Trinity, and Neo, where the computers had enslaved humans to produce energy, while at the same time keeping their minds occupied in a computer-generated world.

The success of *The Matrix* raised the bar for telling stories about simulated worlds and the people living in them. Post-*Matrix* stories had to have deep roots in science and technology to explain how their subjects were integrated into their simulated worlds. Culturally, ideas of simulation theory became part of the vernacular, simulated worlds became just another science fiction convention, and creators found they could spend less time explaining the concept and more time developing the world and its characters.

More sophisticated examples would include the *Black Mirror* episode "San Junipero," which won two Primetime Emmy Awards; "Extremis" in season 10 of *Doctor Who*, wherein a simulated Doctor emailed the real Doctor, who was outside the simulated Earth; and the final six episodes of season four of Marvel's *Agents of S.H.I.E.L.D.* Author Richard K. Morgan put a dark spin on the idea in his novel *Altered Carbon* in 2002 (turned into a Netflix series in 2018), in which a simulated world was a given fact of everyday life, and occasionally used as a place where people—or their minds, at least—could be tortured for any time period desired by adjusting the rate by which time passed.

And of course, those three episodes of *Rick and Morty* (so far).

BOSTROM GOES BOOM

As simulation theory was making itself one of the more well-used tools of science fiction in the early 2000s, Oxford philosopher Nick Bostrom's 2003 paper "Are You Living in a Computer Simulation?" in the journal *Philosophical Quarterly* forced scientists to take the idea more seriously. Bostrom's paper created a mini wave of pop-culture articles and papers and opinions from other academics, and is also responsible for the idea's being discussed and mused upon publicly by celebrity intellectuals such as Dr. Neil deGrasse Tyson, Elon Musk, Max Tegmark, Lisa Randall, and others.

Bostrom's paper (which is available online and not that difficult a read, for an academic text) took what used to be a topic for discussion for stoners and gave it philosophical steroids. The paper is most famous for Bostrom's three-way "trilemma," where three possibilities exist:

1) "The fraction of human-level civilizations that reach a posthuman stage (that is, one capable of running high-fidelity ancestor simulations) is very close to zero," or

2) "The fraction of posthuman civilizations that are interested in running ancestor simulations is very close to zero," or

3) "The fraction of all people with our kind of experiences that are living in a simulation is very close to one."

At least one is true, Bostrom says, which means his three possibilities really boil down to two: 1) either simulations that are indistinguishable from reality will never exist, or 2) we are living in a simulation right now.

Let that sink in.

Now, for Bostrom's ideas to work, one must assume that human consciousness is substrate-independent. This means that human consciousness and the brain are not inseparable, and one does not depend upon the other. A consciousness that we would recognize as human can exist—and perhaps can even be created—

using something other than a biological substrate made up of neural networks and supportive tissues—most likely silicon chips. Basically, this part of Bostrom's argument moves us away from a brains-or-bodies-in-jars simulated world like that of *The Matrix* or a world that is constantly created for you while you move on a treadmill, like the Holodeck or the Zigerion approach.

You don't have to be a card-carrying philosopher to start reasoning through Bostrom's propositions. From the start, you can't rule out #1, as it's become obvious since the Cold War that humanity has the means to destroy itself several times over, and we all too often flirt with the idea of doing so. Proposition #2 can be disproven—that is, if we survive as a species, our far-future descendants will create simulations—based on, if nothing else, the abundance of simulation-style video games currently available for gamers and the advancement that realism saw in just a decade. That leaves us considering #3 as a reality.

And it gets weirder. Within option #3, there are two more possibilities—every posthuman "person" who can run a simulation runs at least one, resulting in billions of individual simulations being run. Or the starts of the stack of turtles where things go recursive—the simulated beings inside the simulations—are so advanced that they create and run their own simulations, and those simulations run their own simulations. It's like zooming in on *SimCity* to find the Sims playing *SimCity* with its own Sims who are playing *SimCity*, or the *Pip-Boy* games of *Fallout 4*, where as a player in the game, you find games within the game which were created by unnamed "individuals" that exist only within the game.

And weirder yet—can the simulations run simulations ad infinitum? If they can, how far is any given simulation from the original society that created it? If there are billions of simulations, and each simulation's simulated "people" can run their own simulation, then the chances that we are (or even that we're anywhere near) the "root" simulation are beyond remote.

To pick a celebrity champion, that's Elon Musk's view. Musk has stated publicly his belief that we are almost certainly living in a simulation, and moreover that we're nowhere near the original simulation, which would have been created by the posthuman life-forms. He thinks we're billions upon billions of simulations away from the base simulation—most likely a leaf simulation rather than a trunk or even a branch simulation.

At Recode's annual Code Conference in 2016, Musk said:

So given that we're clearly on a trajectory to have games that are indistinguishable from reality, and those games could be played on any set-top box or on a PC or whatever, and there would probably be billions of such computers or set-top boxes, it would seem to follow that the odds that we're in base reality is one in billions.

Neil deGrasse Tyson explained it to Larry King in 2017 when the host asked him if he agreed with Musk:

I find it hard to argue against that possibility. You look at our computing power today and you say, "I have the power to program a world inside of a computer." Imagine in the future where you have even more power than that, and you can create characters that have free will—or their own perception of free will. So, this is a world, and I program in the laws that govern that world. That world will have its own laws of physics, of chemistry, and of biology.

Now you're a character in that world and you think you have free will and you say, "I want to invent a computer," so you do. "Hey, I want to create a world in my computer." And then that world creates a world in its computer, and then you have simulations all the way down. So now you lay out all these universes and throw a dart. Which of these universes are you most likely to hit? The original one that started it, or the countless simulations—the daughter simulations—that unfolded thereafter? You're going to hit one of the simulations.

Based on those kinds of statistics, Tyson says, it's hard to argue against the possibility that we're all not just the creation of some kid in their parent's basement, programming up something for their own entertainment. Disruptive events, Tyson says, could be something that the programmer throws in when they get bored and want to stir the pot.

All of that, though? Not science. That's all philosophy.

SIMULATION: SCIENCE VS. PHILOSOPHY

Try as it might, simulation theory has an extremely difficult time getting out of philosophy and being viewed as science. Sure, the technology needed to produce a simulated world reeks of science and scientific development beyond our most fantastic projects, but at its core, skeptics point out, the theory is solid philosophy.

And even the theory's most ardent proponents must agree on some points. While Bostrom did add some formulas and numbers to his paper, his calculations were speculative, and his whole view was from the philosophical side. Why doesn't simulation theory fit into science, despite the lovely word "theory" that science loves to claim as its own exclusive property? Let's talk about what science is.

Science is a way of way of acquiring knowledge about the world around us, whether it's the real world or a simulation. While there are accidental discoveries, true science—the kind that results in a net gain of knowledge—is a process. The steps of science follow a general path that looks like this:

a) An observation is made that leads to a question. For example, seeing that the crops did not grow well this season leads to the question, "Why was the crop yield so low this year?"

b) The question gives rise to a hypothesis. A hypothesis is a proposed explanation for the observation, which can be simple or complex, but it must be testable. The scientific method calls for two competing hypotheses: the null hypothesis, which the scientist is trying to prove wrong, that the observed phenomena occurred by chance, and the proposed hypothesis, which is referred to as the alternative hypothesis—the outcome the researcher predicts. Continuing with our farm example, a hypothesis could be as simple the crops did not grow well because they did not receive fertilizer. The null hypothesis would be that fertilizer has nothing to do with the crop yield, and the alternative hypothesis would be that fertilizer has a demonstrable effect on crop growth and yield. Also, importantly, the proposed hypothesis must be falsifiable—that is, it must be able to be proven wrong. Again, with our farm, if experiments show that giving the crops fertilizer increases yield, then not giving fertilizer must result in reduced yield.

c) A prediction is made based on the hypothesis. What's the outcome of the hypothesis? What should the results of the experiment look like? Again, with the farm, the researcher would hypothesize that giving the crops fertilizer will result in a larger crop yield.

d) The hypothesis is tested via experimentation. The experiments must be designed to specifically test the hypothesis, not to fish around and gather data that has no meaning in relation to it. The more well-designed the experiments to prove the hypothesis is true, the better. For the farm example, plants would be grown with and without fertilizer, with the expectation being that plants grown with fertilizer would thrive while those without would not.

e) The results of the experiment are analyzed and conclusions are drawn. Do the results support the prediction of the proposed hypothesis, or do they support the null hypothesis? For our farm example, if, as we predicted, the crops with fertilizer thrived more than the plants without, then the results support the proposed hypothesis. If the treated plants didn't thrive (or, conversely, if the non-treated plants also thrived), then it's back to the drawing board. The null hypothesis was true.

The thing with simulation theory is that it's hard to make it work within the methods of science. Here's why:

What's the observation that leads to a hypothesis?

In our world, there are no replicable observations about greater reality that would lead to a question, and thus to a hypothesis. No consistent, unexplainable "glitch in the Matrix" has been found. This covers all "supernatural" phenomena that some proponents point to, claiming that anything that's unexplained is proof of a simulation (or any other wild theory). These observations are not repeated and do not occur under controlled circumstances; thus, they don't meet the criteria for a scientific observation.

Contrast this with what Rick sees in "M. Night Shaym-Aliens!": at the start of the episode, he's already spotted sloppy craftsmanship in the rat's internal structure, Morty's responsiveness to Beth's weirdness, and the weather. Rick has had previous experience with the Zigerion designers of the simulation, so he knows what to look for, and as a result, the collective inconsistencies are enough to convince him of the truth. And the glitches only grow in number as the episode continues, and Rick can spot them every time.

We can't. In our world, there are no "glitches" that we know of, nothing that suggests that our world is merely a construct. Nothing to get the science ball rolling.

What's the hypothesis or the prediction?

In the simulation created by the Zigerions, Rick knew that if he overtaxed the computer on which the simulation was running, for example with the rap concert and the string of increasingly complex instructions for the sims in attendance, the computer would freeze. And it did.

But in our world, simulation theory has no such predictions that are testable. No hypotheses can be made. We can't give a large crowd an insanely complicated series of instructions and wait for them to freeze. Likewise, we can't ask people to take off their clothes and fold themselves twelve times or turn coffee in a mug into a farting butt. Saying "Computer, freeze program" out loud just makes people around you think you're weird. To date, reality has not stopped for anyone.

How can it be falsified?

For the simulation theory to be considered science, it needs to be able to be proven wrong. With this theory, we've got two very broad possibilities: either everything is simulated, including any results to any experiments we might run to show that we're living in a simulation; or everything we experience and know is the real universe, and all attempts to show otherwise will give us results that we cannot separate from those that might show we're in a simulation.

In other words, we just can't tell. Ever. It's out of our reach of understanding or proving. We can't test or know this, because we cannot possibly understand it. Following this line of reasoning, you can throw out the simulation theory altogether as a view of reality. We might just be living in a dream. Or in a bead that's part of a cat's collar.

In the end, big-picture, nature-of-existence-type hypotheses often put themselves into untestable categories, and are often causes championed by individuals claiming that science is suppressing information, "big science" isn't designed to see what is actually "there," or that, as a species, we need an entire change in cognition and consciousness in order to understand the true nature of the world. If it all starts to sound like someone is taking their thumb and smudging the line between science and religion, well . . . yeah.

But in the end, these are just things that make scientists grunt and say, "Huh; that's interesting," before they go back to doing science.

SCIENCE STRIKES BACK

Okay—let's go down the science-ish road a little bit.

Despite the scientific problems baked into simulation theory, scientists haven't stopped taking hold of the idea and trying to root it in the world of hypotheses and experiments. The first problem to tackle is, Could a computer that could simulate everything a human being experiences, let alone every living human being's experience, even be built? At least, in theory?

Short answer—probably.

Computers today are vastly more powerful than computers were just five years ago. This progression of increasing computer processing power over time was described by Intel cofounder Gordon Moore in 1965 and quickly became known as Moore's Law, one of the gospels to which computer chip manufacturers and futurists cling. According to Moore's initial reasoning, the number of transistors that we can put on a computer chip would double approximately every 12 to 18 months, while the cost of the chip was cut in half. Over the years, the timing and specifics of Moore's Law have been rejiggered to doubling and halving roughly every two and a half to three years, but it was a fairly accurate approximation of the growth of processing power for decades.

But like all good things . . .

Things can only get so small before physics (or, perhaps more precisely, nanophysics) conspires against you. Currently, the smallest single transistors are 7 nanometers, with 5 nanometers being the next goal. But getting transistors down to 7 (and 5) nanometers costs billions, both in new manufacturing and new silicon technology. Just to scale this out, DNA is approximately 2.5 nm in diameter. A gold atom is about .33 nm in diameter. This is small.

Sooner or later, physics and chemistry—although, at this scale, the two become hard to distinguish—just won't allow things to get smaller and still function as we need them to.

So, Moore's Law is going to wind down, probably around the mid-2020s. In 2015, Moore himself said that he saw it dying out in around ten years. Without big developments in technology, 7 nm may be the end of conventional silicon technology.

Does that mean we'll never make computers more powerful than those that will come out in, say, 2025?

Not at all. Never underestimate human ingenuity in making new technology, either "last-gasp style" with current materials or entirely new approaches that revolutionize the computer industry, such as quantum computing or new directions in cloud-based computing.

Even without invoking unknown future tech and developments in physics that we can't yet fathom, we can reasonably predict the highest processing power we'll be able to develop, which would be the lower limit of the posthuman civilization. And we've had these ideas for a while.

In 1992, nanotechnology expert Eric Drexler outlined how a computer system roughly the size of a sugar cube could be produced that would be able to perform 10^{21} operations per second. Scale up the size of that type of system from a sugar cube to a large planet (a so-called Jupiter brain), and your supercomputer could perform 10^{42} operations per second. New technology, of course, would push those exponents higher, but how much power would the computer need to make a convincing simulation?

Clearly more than the Series 9000 the Federation used with

Rick in prison, and way more than the Zigerion's glitchy processor, since Rick was able to see right through its simulations.

Estimations put the total number of operations per second of a human brain somewhere between 10^{14} and 10^{17} or more, but there are probably redundancies built into our nervous systems that could be left out in the name of increasing the efficiency of simulated people.

Given the numbers, could a planet-sized computer capable of 10^{42} operations per second simulate all the people who currently live on Earth? Let's do the math.

(I'm not going to calculate all the people on Earth ever, because if we're thinking about simulation theory being real, let's go all in. I'm not sure the past even happened. Maybe the past is a memory implant or part of the programming. That said, I do know I'm experiencing things right now, and I assume you are too. Let's look at the simulation being able to create the reality of the year 2100. If the computer has enough power to simulate 2100, it can easily handle simulating now.)

How many people should the computer simulate? Let's use 11 billion—that's the projected population of the Earth in 2100, which will definitely cover our current 7.7 billion or so.

We can ballpark the average human life span at seventy-five years. It's a little higher in developed regions, a little lower in less-developed regions, and should only increase as the years go by (yay, science!). Oh, and since we'll need this, there are 3.2×10^7 seconds in a year (365 days x 24 hours x 60 minutes x 60 seconds).

And, being generous, each of those 11 billion simulated brains must experience somewhere between 10^{14} and 10^{17} operations per second to convince them that they are in a physical reality.

So:

1.1×10^{10} x 75 years/human x 3.2×10^7 seconds/year x 10^{14} to 10^{17} operations/second =

2.64×10^{33} to 2.64×10^{36} operations per second needed to simulate reality for 11 billion people.

(A note for naysayers: increase the number of people way up, decrease the life expectancy down, or the jiggle the number of operations per second and you'll come up with the same range of exponents.)

The planet computer can perform 10^{42} operations per second, so how much of the planet computer's processing power is that?

$$2.64 \times 10^{33} \text{ to } 2.64 \times 10^{36} / 10^{42} = 2.64 \times 10^{-9} \text{ to } 2.64 \times 10^{-6}$$

To simulate reality for 11 billion people would use only a millionth to a billionth of the planet computer's total processing power. Heck, you wouldn't even need a large-planet-sized computer. A moon-sized one might do the job just as well.

But remember—this isn't to make a simulated universe, it's the computing power needed to make the human mind believe that it's in a simulated universe.

So, yeah, it could, in theory, work.

Someday.

BUT ABOUT THAT SCIENTIFIC PROOF...

Even though a computer with the power needed to simulate reality for 11 billion people could, in theory, be built in the future, that doesn't mean science is saying that simulation theory is a go. Skeptics' claim that you can't treat simulation theory like a science because there are no observations to be made still stands. While more and more scientists are saying that we may live in a simulation, scientists with legitimate expertise in the field aren't exactly numerous, and there are plenty who aren't convinced by any of the claims.

Many of the claims of "proof" aren't the kind of observations that science requires. Jump into even the shallow end of YouTube about this, and you'll be presented with dozens of "Simulation Theory: Now We Have Proof!" videos. Most offer up interesting scientific phenomena that, when explained in the context of living in a simulation that's programmed by a simulator, can sound convincing. But they're not proof.

For example, Rick pointed out a Pop-Tart living in a toaster and driving to work in a smaller toaster car to Morty. That observation—and the fact that they could see it each day—doesn't prove that they were in a simulation. It was just something weird. Okay, something really weird. Only when taken with other evidence, and finally physically running out of the simulation, was Rick's hypothesis confirmed.

It's the same case with many of the "proofs" given by proponents of the idea—and they can get trippy quickly. There are a lot of as-yet-unexplained physics- and math-based phenomena out there, but the scientist's job is to be skeptical about them, not speculative. Science is the journey of taking what is observed and developing reasonable, testable explanations, rather than claiming an untestable explanation as the cause for as-yet-unobserved phenomena, as is the case with those who claim we live in a simulation.

For example, several proponents will point to universal constants—the fine-structure constant, pi, or the Golden Ratio—as evidence of a "code" that underlies our reality. These are "set" values—"untweakable," as far as we know. Why are these values what they are? Could they be part of the source code of this simulated reality, or, perhaps more simply, just things we haven't figured out yet? Science leans on the latter rather than the former.

The other problem with the "proofs" is that they are often misrepresented by proponents and pop news reporters who don't have an understanding of or background in science. For example, physicist James Gates made headlines in the simulation theory arena when he said that, while studying supersymmetry and string theory, he found "error-correcting codes" similar to Shannon coding (a type of data compression) in the equations. Many news and other outlets of dubious scientific credibility amplified and augmented Gates's initial claim, saying that there was a "computer code" in equations that had been discovered that must have been written by someone or something with administrator-level control of our reality.

But there wasn't. Regardless, Gates quickly became a hero of the pro-simulation groups and worked to distance himself from what the claim had become, saying that there are other instances of error-correcting mechanisms in the universe; for example, in all living things. When asked by Neil deGrasse Tyson at the 2016 Isaac Asimov Memorial Debate to put a percentage on the possibility that the universe is a simulation, Gates put the chance at only 1 percent.

All of that is not to say that there isn't any legitimate research going on to test the hypothesis that we're living in a simulation; it's only to say that there's a difference between pro-simulation theorists' "proof" and hard, scientific evidence. Sometimes a cosmological constant is just a cosmological constant.

The physicists who are seriously looking at the idea that our reality may be a simulation are focusing on looking for repeats in the fundamental rules of the universe, suggesting a coder cutting and pasting the same piece of an algorithm over and over or something that could be seen as a "signature" of the code that runs the simulation.

This category of experiments deals with our expectations, based on an infinite universe, not meeting up with what we observe—which, if it was simulated, would be a finite universe. A finite universe is what researchers would expect to find if reality was in fact a simulation, running on a finite computer with finite resources. As an approach, it's like Rick's rap concert. The Zigerions were known for cutting corners, so Rick's test was an easy one. But what about our possible simulators?

Zohreh Davoudi, a physicist at MIT, is working with colleagues on models of the strong nuclear force, which holds the nuclei of atoms together. The models are run on supercomputers that, put simply, are fed the rules by which the subatomic particles act and the virtual space-time in which to act and run. In other words, subatomic particle formation and interactions are simulated, and it takes a tremendous amount of computer power. As technology's capabilities for such simulations increase in the coming years, Davoudi suggests that simulations will be able to move into the macro scale and include cells, humans, and more.

In a very *Terminator 2*, OMG-the-first-Terminator's-hand-was-the-start-of-Skynet-in-the-future-which-sent-the-Terminators-back-to-get-the-Connors twist (minus the evil robots), Davoudi even hypothesizes that the types of simulations she and other researchers in her field are running are the starter technology for full-on universe simulation. As such, she reasons, present-day simulations created on classical computers and simulating the quantum realm of the subatomic will have signatures that will be experimentally detectable—and quite possibly repeated in our own universe, if it is, in fact, a simulation. There's more to be developed between here and there, but at least it's the start of some actual science on the topic.

Davoudi also argues that if we're living in a simulation, then the simulation itself is finite, and thus the laws of physics would exist on a series of finite points in a finite volume. If that's the case, Davoudi says, then high-energy cosmic rays that move at relativistic speeds would behave differently than what we predict, because their motion started sometime and they came from somewhere in a finite space of the simulation.

But while Davoudi's experimental ideas seem to be good science, the technology needed to perform them isn't quite ready yet.

WHERE DOES THE RABBIT HOLE END?

For the coming years, probably with philosophy. Science just doesn't have the tools to approach it yet. Period.

But does this mean that we are or we aren't living in a simulation? That what we know as the entire history of the species is—and this is one of Tyson's favorite ways to explain it—maybe an evening's entertainment (or perhaps a game called "Roy") for some posthuman adolescent waiting for their mom to cook dinner?

Yes.

No.

Whatever you want to think.

All the answers are correct. All the answers are incorrect. And that's okay.

Maybe, if we *are* in a simulation, finding out that we're in one is the endpoint, the moment in the game at which the simulator shuts it down or restarts it. Maybe, if we're in a simulation, we're programmed to never be able to figure out that we're in a simulation. Or maybe, if we are, the point is to lead such imaginative, amazing, exciting lives full of creating and doing things that the simulator will always be interested and never want to shut it down.

Or maybe we're not, and that's okay too.

To quote Douglas Adams, "Isn't it enough to see that a garden is beautiful without having to believe that there are fairies at the bottom of it, too?"

Or, if not fairies, little bits of code that look like fairies to our simulated brains.

CHAPTER 19

Technology

★ ★ ✦ ★ ★

Technology and science are often used interchangeably, and certainly Rick works in both areas. But as we focus on the technology of *Rick and Morty*, we should probably draw some lines separating the two into their distinct areas.

Science is a method, a systematic process of acquiring a body of knowledge based on observation, experimentation, and prediction. Science can be broadly divided into two categories: basic, which is knowledge for knowledge's sake, and applied, which is the application of the knowledge gained by science to specific needs.

Technology is broader. The word comes from the Greek *tekhnologia*, which means the systematic treatment of an art, craft, or technique. Put simply, technology is the application of scientific knowledge for practical purposes or in order to solve problems. Technology, as we've come to know it, can be a piece of equipment, a collection of techniques, or an activity that can either form or change a culture.

Science and technology are dependent upon each other and overlap as scientists, designers, and makers of technological innovations are always moving back and forth, into and out of the others' camps. But technology as a concept is often seen as the product of science. Science learns new information, and technology takes it and sets it free. The cycle can go in reverse, of course, as new technology can lead to new scientific discoveries, and on and on.

Rick covers both bases. While his adventures seeking out new knowledge are not showcased, he must have done it, and most likely is continually looking to learn new things about the universe. Meanwhile, the garage and the labs are littered with products and items that are complete or in some state of assembly. Quick glimpses of plans are seen here and there. And, of course, Rick's technology has been shown to change and evolve throughout the course of the series. Consider the first version of the Portal Gun seen in flashback in "The Rickshank Rickdemption" with the version he uses in present day.

Our technological advancement as a species is relatively short, if you look at our history. Prehumans used stone tools for millions of years, and we can put the emergence of more specialized stone tools, such as awls, spear points, and hand axes, at around 200,000 years ago. Further development was slow. Controlled fire, food, and shelter were all early hits for us as a species, as were the use of metals and the harnessing of kinds of energy other than fire, sun, and human muscle. Each new development continued to evolve as new information was learned and applied to the technology. And from it, we got better wheels, roads, plumbing, and much more.

Throughout history, advances in technology have occurred without formal science behind them, mostly through trial and error. The Egyptians and the Romans knew how to make glass; the swordmakers of Toledo, Spain, could make the finest swords; indigenous peoples knew how to make boats that could take them thousands of miles; but none fully understood the science behind their respective activity.

Additionally, for centuries, there were limitations on the progress of both science and technology, as information was lost and rulers (both secular and religious) tried to place limits on what man should be allowed to know or in what directions science could look. It took us until the Renaissance to really get going in terms of science and technology.

Over time, some controls and restraints loosened, and science and technology became more intertwined. Each pushed the other further ahead, and often pulled civilization along behind it. The sheer scientific, engineering, and technological momentum of the Renaissance led directly to the Industrial Revolution in the late eighteenth century, which led to the Technological Revolution of the nineteenth and twentieth centuries. And the modern age of innovation has led us to this point today.

But the pace at which science and technology move forward changes due to a multitude of reasons. Science is most often driven by curiosity, while technology is driven by demand (consumer, industrial, war, etc.). As science delivered new knowledge throughout the 1800s, for example, industry took the knowledge and developed the associated technology that made use of it. Michael Faraday, for example, discovered the principles behind the generation of electricity, but it was French instrument maker Hippolyte Pixii who developed the first generators, based on Faraday's ideas.

But by the beginning of the First World War, science was commandeered to serve the military, and resources were focused. Science moved forward quickly, but toward a specific wartime goal.

Technology continued to race ahead throughout the twentieth century, but with science that was often focused on specific problems like the Second World War, the atomic bomb, putting a man on the moon, and the Cold War. Rather than being broad-based, technology often took what we had and made it better, largely in the service of capitalism. We had cars, then we had better cars, and better-better cars, and better-better-better cars, and on and on and on while the science stood still. But now that the demand exists for electric cars, it has changed.

Again, technology needs science and science needs technology. Each can survive on its own, but it's not a recipe for moving things forward. Without new discoveries, technology just makes the same thing over and over with better bells and better whistles.

TECHNOLOGY, IT'S A-CHANGIN' FAST

Technology moves forward, and based on how we've experienced it in recent history, the rate of that forward motion tends to increase. While it took over a hundred years from the Industrial Revolution to the first airplane flight in 1903, it took less than a human lifetime after that to land men on the moon in 1969. Think how many different cell phones you've accumulated in just the past ten years. Technology changes. More than just changes, it evolves.

Showing similar traits to biological evolution, technology changes over time as the result of small but significant alterations. Given enough time, these changes will result in a new "species," or, in this case, a new piece of technology that's barely like any that came before. Compare an iPhone to an early portable "bag phone," or, worse yet, a car phone that had to stay in the car.

There are many approaches to predicting technology's rate of change over time, with perhaps the best being the previously mentioned Moore's Law. While Moore meant for his formula to be applied only to describe changes in computers, it has now been generalized to apply to any technology to state that the rate of improvement in any specific tech will increase exponentially over time. The specific rate depends upon the technology in question.

While changes in computing technology are pushing the limits of Moore's Law, research on past trends shows that the law works quite well to describe how rapidly technology advances—and, as stated, it's not linear but exponential. Futurist Ray Kurzweil uses a similar paradigm to describe changes in technology called the

"Law of Accelerating Returns," which supports the exponential nature of Moore's Law and adds in the inventions of new technology to help cross barriers to current tech. Change in technology, according to Kurzweil, will only come faster and be more profound and life-changing as time goes on. The rise of the Internet in the mid-1990s was the last one. Ask anyone who was around to describe life pre-Internet, and it sounds like another world.

Imagine for a minute that you've jumped back to 1995. Show the people living in that dark age (who think that Netscape is the be-all and end-all) an iPhone. You're a wizard. Your audience can possibly see the connections to their technology, but still, a lot of it is just magic.

Now, since you can travel through time, jump ahead twenty-five or so years from the present day and take your snazzy, modern iPhone again. Given that the rate of change is accelerating, it's likely that the future's technology will make your modern iPhone look like a child's toy.

SCIENTIST RICK

As mentioned earlier, Rick is a scientist. He's talked about collaborating with others from time to time, but in his heart, he's a loner, and therefore he must do virtually all his science on his own. His discoveries then inform and fuel his technological advances. By the time we meet Rick, both his science and his technological advances—what he knows and what he can do with what he knows—are impressive.

The science at the base of much, if not most, of Rick's technology is quantum mechanics. As scientific disciplines go, it's relatively young. Quantum mechanics grew out of atomic theory, the idea that everything is made from atoms that are indivisible and join with each other in small, whole-number combinations like H_2O (among other claims). Out of our understanding of the atom, including the existence of individual atoms proven by Einstein in 1905, we started to comprehend what the smaller, subatomic particles could do. And it was odd.

Quantum mechanics has three main, albeit weird, ideas: that light is made up of particles called photons; that all matter exhibits wavelike behavior; and that all light and matter have a baked-in angular momentum with specific, discrete values. This understanding of matter allowed for the creation of a vast range of technology, including semiconductors, lasers, compact discs, transistors, cell phones, fiber optics, GPS, atomic clocks, MRIs, and many more of the inventions that fill modern life. It also explains how the basic building blocks of reality work.

The progress that quantum mechanics–based technology has made is 1) astounding, and 2) solidly based on the science of the very, very small. Again, the marriage between science and technology is strong.

The promise of quantum mechanics is equally astounding, and we're already nipping at the heels of some of the most promising technologies, including quantum computers, which would be vastly more powerful than modern computers and would be based on an entirely different method of decision-making and storing information. Quantum computers will be able to handle incredibly complex problems in an exponentially shorter time period than regular computers by performing far, far fewer operations than classical computers would.

The full integration of quantum computers and society will be another fundamental change, akin to the effect of the Internet, but bigger. Much bigger. But it's likely still years, if not decades, away.

To be able to do what he does, Rick has to have a full understanding of quantum computers and quantum physics, as well as a mastery over other disciplines such as genetic manipulation and tissue engineering (to successfully grow clones), astrophysics (to understand wormholes), and energy production and management (to power his tech), as well as all the fields that support them. So let's try to figure out the pace of Rick's technological advances and compare that to the rate of change of the world.

In our world, it took about fifty years for quantum mechanics to go from a collection of weird ideas to being used in the production of semiconductors, which fundamentally changed the world. Fifty more years and we're virtually in the present day, as far as quantum physics–based technology goes. So, it took the world—and that's thousands of scientists (and designers, engineers, and manufacturers) working on the problems—one hundred years to get us from ideas to here. Not that bad, considering the technological progress of the previous hundred years. The rate does seem to pick up a bit.

Now, to Rick. One guy working on his own for the most part, on multiple disciplines but focused on the quantum mechanics that will ultimately lead to his creation of portal technology (which in and of itself utilizes several other types of tech than just quantum). To ballpark a time frame, let's assume there was a grain of truth in Rick's memories as seen in "The Rickshank Rickdemption" and that Rick made his breakthrough in portal technology when Beth was about six years old. She's around thirty-four in the present, so that means Rick had about twenty-eight years to go from very smart guy who was already ahead of anyone else when it came to quantum mechanics to the Rick Sanchez that we know and love.

Quick sidenote: Rick was in a rock band, the Flesh Curtains, with Squanchy and Birdperson when he was in college, but let's assume that he was hitching rides with his intergalactic bandmates.

So—again, just using this to ballpark here—if the scene in "Rickshank" is at least a little accurate, he was about to make his breakthrough around 1990. He's had about twenty-eight years to do what he's done. And what he's done far outstrips the pace of science and technology in our world. For Rick to have learned everything he knows on his own and then develop it all into working technology, his rate of change would be ten to one hundred times greater than ours—and remember, ours is already looking to be exponential.

Yes, some of this is plot-based to make the stories move, but effectively—and he's even said it outright at different times—Rick knows everything there is to know. Not in the hyperbolic sense; in the sense that he functionally knows everything. Like, *everything* everything. He turned himself into a pickle, after all, and by our known science that's just, well, magic. Rick knows all and can apply all.

It sounds cool, but that's not good. Here's why . . .

THE SANCHEZ STAGNATION, THE MORTY SALVATION

One of the worst things that can happen to a civilization is that its rate of scientific and technological growth slows and stagnates. Take away science, and, as mentioned earlier, technology can continue to move forward but will eventually reach a point where making a gun that's bigger than the last one is the only innovation it can offer.

Rick knows everything, but that's a curse, not something cool.

There's nothing left to learn; there's no direction for him to grow in.

Sure, Rick can make anything he wants in order to do anything he wants, but nothing is ever going to be "new" again. This is it. There's nothing left for Rick. If anything, this explains, or adds to, his ennui, that existential disgruntledness that he seems to exude throughout the series. It's a mood that Rick has certainly sunk into many times, and perhaps found himself dealing with regularly prior to the series' start.

Maybe this is why Rick came home to the Smiths. By himself, there was nothing left. At least with Morty, there's the introduction of uncertainty. A chance to see the universe to which he's become so jaded through new eyes. A little chaos to keep things from getting too predictable.

Acknowledgments

* * * * *

This book would not exist if it wasn't for a blind email sent by one Oliver Holden-Rea asking if I might like to talk about writing a book about the science in *Rick and Morty*. He convinced me that he was a) a real person, and b) not joking, and has been nothing but the best editor a noob could hope for throughout this entire process. And thanks go to Charlie Beckerman as well. Together, you guys took my caveman-speak and turned it into what it is here. If there are any mistakes in here, they're all mine, and I alone deserve the blame. Oli and Charlie are princes among men.

Also, thanks to the following people who so generously gave their time to answer emails or take phone calls from someone asking seriously weird questions about their disciplines: Jonathan Losos, Yelena Bernadskaya, Daniel Korostyshevsky, Sophia Nasr, Spiros Michalakis, and Richard Matzner. You were all great sports.

I'd also like to give a nod to the work of scientists and science communicators who inspired me both over the years and while writing this book: Brian Greene, Kip Thorne, Lisa Randall, Max Tegmark, Sean Carroll, E. Paul Zehr, Rhett Allain, Neil deGrasse Tyson, Suveen Mathaudhu, Hank Green, Adam Frank, Sam Kean, Patrick Johnson, Ethan Siegel, Derek Muller, Chad Orzel, Katie Mack, Michio Kaku, and Brian Cox.

And while we're talking inspirations and acknowledgments, decades before any of this, there was a little kid watching Carl Sagan present *Cosmos*. There was a late teen watching Bill Nye talk enthusiastically about science. And there was a grown-up going into teaching science watching Adam Savage embrace his passion for making and learning on *MythBusters*. And then, there was a science teacher watching any video of Richard Feynman he could get his hands on and soaking up the knowledge and the sheer exuberance on display.

All four had a hand in making me who I am when it comes to teaching and communicating science. Forever indebted.

And finally, thanks to Dan Harmon and Justin Roiland for your wonderfully twisted creation, in all its weird, irreverent majesty. Here's to years and years more. Please don't name a character after me and kill him off in a horrible way.